HISTORIAN'S GUIDE TO STATISTICS

QUANTITATIVE ANALYSIS
AND HISTORICAL RESEARCH

HISTORIAN'S GUIDE TO STATISTICS

QUANTITATIVE ANALYSIS AND HISTORICAL RESEARCH

CHARLES M. DOLLAR
Oklahoma State University

RICHARD J. JENSEN
University of Illinois
Chicago Circle

HOLT, RINEHART AND WINSTON, INC.

New York   Chicago   San Francisco   Atlanta
Dallas   Montreal   Toronto   London   Sydney

To Hanne and Martha

PREFACE

This book is a practical guide to the use of quantitative methods and computers in historical research. We have not assumed that our readers are trained in mathematics, statistics, or computer science. Our goal is to provide a step-by-step introduction to all the important techniques necessary for conducting serious quantitative research using historical data. Consequently, the material may be of interest not only to professional historians and graduate students in history, but also to social scientists who have occasion to analyze quantitative data from the past. And by quantitative data we mean documentary remains of the past. For this reason we do not discuss questionnaire design and survey research, the most popular form of "experimental" social science.

The book is divided into four sections, each of which can be read independently of the rest. Chapter One is an exposition of the purpose and methods of quantitative research. Chapters Two, Three, and Four cover the statistical techniques of greatest value in dealing with historical data. Chapters Five and Six discuss the principles of electronic data processing (EDP) and their application in historical research. The last chapter brings together a comprehensive annotated listing of the most useful sources of quantitative historical data and the most relevant theories and methods of the social sciences.

The intellectual debts we owe our fellow quantifiers (and our occasionally skeptical colleagues) is enormous. Among the many scholars who have helped us, we must especially thank Thomas Alexander, Howard Allen, Lee Benson, Al Bogue, Brad Burke, William Chambers, Thomas D. Clark, Jerome Clubb, Samuel P. Hays, James Henderson, Paul Kleppner, Warren Miller, Murray Murphey, Roy E. Schreiber, Joel Silbey, John Sprague, Stephen Thernstrom, S. Sidney Ulmer, Sam Bass Warner, C. Vann Woodward, Robert Zemsky, our students, and Inter-University Consortium for Political Research students.

We gratefully acknowledge invaluable support from the International Business Machine Company, the Inter-University Consortium for Political Research, the National Science Foundation, the Oklahoma State University Research Foundation, the Oklahoma State University Computer Center, the University of Kentucky Computer Center, the University of Michigan, Washington University (St. Louis), and Yale University.

Responsibility for the manuscript rests jointly with the authors; with Jensen drafting chapters Two, Three, Four, and Seven, and Dollar drafting chapters One, Five, and Six.

Stillwater, Oklahoma   C. M. D.

Chicago, Illinois   R. J. J.

November 1970

CONTENTS

HISTORIAN'S GUIDE TO STATISTICS

QUANTITATIVE ANALYSIS
AND HISTORICAL RESEARCH

INTRODUCTION

C HAPTER O NE   Two misconceptions threaten to
                      impede the use of quantitative
                      methods by historians: fears of
dehumanized history and mistrust of an alien methodology.

The notion that quantitative methods will produce arid, mathematical, deterministic history is unwarranted. Historians who fear that quantification will smother the "human element" and "qualitative factors" in history have not to our knowledge published a detailed statement of their criticisms; conversely, no historian to our knowledge has maintained that quantification is the new dispensation that will unlock all the secrets of the past and reduce the corpus of traditional historiography to so much scrap paper. The bogey of deterministic history that ignores the "human factor" and somehow causes people to behave the way they did should not trouble the historian.

The criticism that quantitative research displaces insight and imagination with mechanical routine is unlikely to occur to a historian who has actually attempted explicit quantification. The actual coding of data and the punching of IBM cards is, indeed, routine work—much like typing and note taking. However, it is hardly the essence of quantification. Much more important is the design of a research strategy, the adaptation of codes to capture the essence of the data, and the actual analysis and re-analysis of tables. None of this work can be performed adequately by anyone unfamiliar with the historical context of the data and the subtleties of historiography. Many classic quantitative research designs have, indeed, involved months and years of painstaking work, only part of which was handled by assistants. The traditions of the profession accord high esteem to historians who engage in meticulous, painstaking, and time-consuming research in the sources; we can not believe that the advent of new methods will alter this heritage. There is no royal road to historical insight, but we believe quantitative methods can often speed the trip.

Another criticism is that a total reliance on quantitative data obscures the critical importance of "qualitative" data and hence inhibits a grasp of the true context of the past. Much of the criticism of the new economic history is a reaction against excessive reliance upon economic statistics and an avoidance of social and political factors. Social and political historians have never to our knowledge suggested that statistics tell all. Furthermore, it would seem that the dichotomy between "quantitative" and "qualitative" data is largely a false distinction. Practically any attribute that people, events, or institutions possess can be quantified in some way or another. It is true that when the data are of poor quality because of obscurity or meagerness, quantification will not work; nothing else will work well either.

Few historians will find statistics useful in the analysis of a single idea, value, or belief, or in tracing the logical interaction of ideas or their psychological implications for one man. Quantitative methods may, however, help analyze the distribution of ideas among men of different experiences and interests, and the distribution of support for an idea across time. The question of what kind of men accepted, rejected, distorted, or ignored a particular idea is often of great importance, and, if enough men are involved to make explicit quantification worth the trouble, the historian will find a large body of useful statistics and data processing techniques available. Values and emotions are equally amenable to statistical methods, so long as they can be precisely attributed to the men under study. The historian who talks of "widespread" fears or "growing" insecurity has subjected his data to implicit quantification. If his sources are adequate enough, he should have no qualms about explicit quantification. True, the "spirit of the age" and the "climate of opinion" that a historian uses to explain events may be indicated by so many diverse and fleeting bits of evidence that no systematic procedure to collect, measure, and compare the data is possible. We trust that in such situations the researcher will display common historical sense and not reject evidence that cannot be neatly quantified.

When quantification is called for, it must be done carefully and explicitly. Perhaps the greatest advantage accruing to explicit quantification is the economical processing of large quantities of data. The historian confronting a mass of election statistics, or plantation account books, or colonial wills, or a long list of names can materially lighten his research burden by using statistics and computers. If the mass of data is too great, he can use sampling techniques to select a reasonably small body that will yield patterns very similar to those in the original corpus. If he is interested in one attribute, like income, he will find that simple descriptive statistics exist which permit him to boil a mass of numbers down to a convenient handful, such as the median, the mean, or the index of inequality. If two or more attributes are studied for interrelationships, he will want to use correlation and regression coefficients that summarize in a few numbers the strength and form of relationship. If his data refer to consecutive years or decades, he will find it useful to plot time series on a graph. If he is studying the voting behavior of legislators, he will find indices of agreement, cohesion, scale position, and success extremely helpful in making sense out of a mountain of yeas and nays. Content analysis will suggest how the rhetorical "fingerprints" of a number of speeches, sermons, platforms, pamphlets, or editorials can be constructed and compared. Electronic and mechanical data processing techniques are available to sort, count, classify, and compute the requisite statistics, thus saving many hours of painstaking hand work.

Is quantification and scientific methodology an alien approach borrowed indiscriminately from fundamentally anti-historical social sciences? A brief review of the role of quantification in the development of American historiography will indicate that, on the contrary, statistical methods were important in historical analysis even *before* they became important in other social sciences. Frederick Jackson Turner, the most influential American historian of the early twentieth century, repeatedly emphasized the need to establish history among the social sciences. Expressing the consensus of his generation of "scientific" historians, Turner in 1904 proclaimed, "It is safe to say that the problems most important for consideration by historians of America are not those of the narrative of events or of the personality of leaders." After reviewing the need for research in various substantive areas, Turner concluded that the "first problem" in the discipline was nothing less than "the problem of how to apportion the field of American history among the social sciences." And he went further:

> No satisfactory understanding of the evolution of this people is possible without calling into cooperation many sciences and methods hitherto but little used by the American historian. Data drawn from studies of literature and art, politics, economics, sociology, psychology, biology, and physiography, all must be used. *The method of the statistician as well as that of the critic of evidence is absolutely essential.*[1]

Turner was addressing himself to the entire range of American history; and his manifesto was *not* exclusively concerned with the frontier thesis. In his seminars at Wisconsin and Harvard he trained scores of outstanding historians to use quantitative research methods. Thanks to Turner, his students, and like-minded colleagues—especially Orin Libby, U. B. Phillips, William Dunning, Charles Beard, E. E. Robinson, Arthur

C. Cole, Carl Becker, Dixon Ryan Fox, Arthur Schlesinger, Sr., Charles Paullin, Joseph Schafer, and Clifford Lord—scores of quantitative studies appeared from 1894 to the 1930s.

The preferred technique in early quantitative history was the statistical map. Borrowing a technique pioneered by geographers in the Census Bureau, Turner taught his students to plot the distribution of election returns, roll call votes, soil types, ethnic groups, and other social, economic, and political variables. Avery Craven recalled:

> One lasting impression which the student carried away from Turner's classes and from his workshop was that of countless maps, jigsaw in appearance, because they represented the plotting of votes by counties. Such graphic representation revealed sectional interests, the force of habit, the persistence of viewpoint carried by the migrants from older areas into newer ones. When thrown against geological survey maps, racial maps, or cultural maps of various kinds, they added something to the American story not to be found elsewhere. Turner gave the United States census maps a new place in the historian's equipment.[2]

Turner's justification for the use of statistical maps derived both from their use as a data processing technique, and from their role in his geographical theory of history. The vast quantities of numerical data available in census reports and other compilations were mere jumbles of digits in their raw form, and virtually impenetrable to historians ignorant of statistics. Once the numbers were converted into rates and percentages and brought together on a map, however, they began to make sense. Statistical regions stood out, often cutting across state lines but usually following natural physiographic boundaries. The visual comparison of two maps, furthermore, permitted a crude sort of correlation between variables. Turner recommended that his students should "make in selected areas, [a] detailed study of the correlations between party votes, by precincts, wards, etc., soils, nationalities and state origins of the voters, assessment rolls, denominational groups, etc. [to discover] what kind of people tend to be Whigs, what Democrats or Abolitionists, or Prohibitionists, etc."[3] Turner was confident that map research would produce significant results because of his fundamental belief in the importance of geography in social development, a belief basic to both his frontier and his sectionalism theories. "There is and always has been a sectional geography in America," he wrote in 1925. "There is a geography of political habit, a geography of opinion, of material interests, of racial stocks, of physical fitness, of social traits, of literature, of the distribution of men of ability, even of religious denominations."[4]

Turner and his followers advanced beyond maps to the important technique of statistical collective biography. Inspired probably by Henry Cabot Lodge's painstaking statistical analysis of all 14,243 names in *Appleton's Cyclopedia of American Biography*,[5] Turner began in the mid-1890s to promote studies of the origins, personal characteristics and constituencies of political leaders. At the same time his extraordinarily original student Orin Libby pioneered the technique of statistical analysis of roll call votes in legislatures.[6]

A good example of early Turnerian collective biography was an unpublished undergraduate honors thesis written under the supervision of Libby and Turner in 1901.[7] The young historian (who later became a wealthy railroad lawyer and lobbyist) tabulated the distribution of party affiliation, place of birth, occupation, and (using the manuscript federal censuses) the personal and constituency wealth of all the members of the Wisconsin legislatures of 1850, 1860 and 1870. He then selected nine critical issues, found corresponding roll call votes, and compared the supporters and opponents of each position in terms of personal wealth, nativity, and occupation.

The Turnerian style of quantitative historiography won professional approval by 1900. The extensive use of statistical maps in the distinguished *American Nation* series (25 volumes, 1902–1906) indicates Turner's success. William Dunning, influenced both by Turner and by several sociologists, taught statistical mapping to his students at Columbia, which emerged as another center of quantitative research. The most important example of collective biography came with Charles Beard's immensely influential study, *An Economic Interpretation of the Constitution* (1913), which together with his *Economic Origins of Jeffersonian Democracy* (1915) relied explicitly on the inspiration of Turner and Libby.

In the 1920s quantification spread to political science,[8] economics, journalism research, and new areas of sociology, but gradually declined in history. After Turner's death in 1932 the only significant historical studies relying upon quantification came from highly theoretical agricultural historians. Leaders among them included Ulrich B. Phillips, Robert Russell, Lewis C. Gray, Avery Craven, Joseph Schafer, Frank Owsley, James Malin, Herbert Weaver, Barnes Lathrop, Paul Gates, Fred Shannon, Fulmer Mood, Lee Benson, Merle Curti, Lawrence Harper, Chase Mooney, Allan Bogue, John Shover, and Thomas Pressly. They kept the spark of quantification and scientific research alive in the profession from the 1920s to the mid-1960s. For the most part, however, their influence was circumscribed by a general decline of interest in rural America. Why did scholars outside of agricultural history abandon a successful method? Did they uncover basic shortcomings or fallacies, or learn better ways to study the past? Some consideration of these questions is essential to an appreciation of the profession's attitude toward quantification.

Historians after the First World War began favoring topics less amenable to quantification since the chief data analysis technique remained statistical cartography. The increasing availability of manuscripts, archival materials, old pamphlets and newspaper files meant that statistical compilations no longer constituted a major fraction of the primary sources open to research. Immigration and labor history, together with diplomatic and intellectual history became more salient, and no one explained how quantification might be useful in this research. The Beardian "economic interpretation," furthermore, suggested a single-factor approach that made multi-factor statistical analysis rather irrelevant. Individual rather than collective biography also became fashionable among academic historians, thanks to the contributions of 2600 scholars to the great *Dictionary of American Biography* (20 volumes, 1928–1936).

On a deeper level, however, the decline of quantification must be attributed to interval flaws in the Turnerian approach itself. Statistical cartography, for all its advantages in geographical interpretations, is generally a poor method of processing numerical data. Good maps are immensely time-consuming to draw and interpret, especially when an amateur cartographer crams his sheets with all sorts of colors, lines, symbols, and other confusing paraphernalia. Even when two variables are in fact correlated, it is hard to discover this by comparing two maps, and it is next to impossible to control for the effects of a third variable. Even when a correlation is spotted, the problem of the ecological fallacy becomes acute—that is, two variables may have similar *spatial* distributions, but may not be correlated at all among individual people. As the following chapters suggest, there are many techniques of data analysis superior to statistical cartography.

The Turnerians were remarkably negligent in not learning about and adopting new statistical methods. Herman Hollerith invented the punched card and the mechanical countersorter in the 1880s, and quickly secured its adoption by census bureaus in Washington and in Europe Except for Frank Owsley and his students who used punch cards and sorting equipment during the 1930s and 1940s, historians generally ignored the use of punch cards until Merle Curti and Bernard Bailyn demonstrated their utility in the 1950s. (Other social sciences showed almost as great a cultural lag—only the popularity of electronic computers in the 1950s and 1960s shook the lethargy of political science, for example.) Turner, despite his fine talk about interdisciplinary cooperation, never approached statisticians to see what they had to offer. Such elementary techniques as random sampling, semi-log graph paper, and correlation coefficients remained wholly unknown to historical research until quite recently.

Worst of all there was a failure of communications among Turnerians, which prevented the rapid and accurate dissemination of the shortcuts, discoveries, and technical advances they discovered individually. No journal published methodological notes; no handbook explained simple techniques; no bibliography guided the neophyte to significant articles or valuable sources. True to the old narrative tradition the Turnerians even removed most of their methodological scaffolding before publishing their statistical research, thus leaving readers confused about exactly how results could be duplicated. Only Turner himself seems to have known about most of the technical advances of his students, and he passed on only a small portion in his seminars. Consequently, there was no cumulative progress in quantitative techniques—indeed, the usual story after 1901 was regression. The historians of the 1950s and 1960s who wanted to learn statistics could rely on no heritage of technique from the past, and had to turn to friends in neighboring social science departments. On the other hand, every historian had read something or other from the old Turnerian school—or at least Beard—and as a result the occasional new articles and books that used quantification did not seem wholly strange. Even the skeptical historian read quantitative work with a sense of *déjà vu*—and perhaps asked himself how he could apply the same techniques to his own research problems.

By the mid-1960s the historical profession included a sizable number of scholars interested in re-establishing the quantitative methodological tradition. Numerous conferences, training programs, data acquisition committees, and regular courses were organized for the purpose of exploring and disseminating the quantitative approach. At the same time the need for a handbook of quantitative methods in historical analysis became increasingly clear. This book attempts to meet this need in two ways. First, it is a bridge between the old and the new—the Turnerian approach and more recent statistical techniques along with the computer. Second, it examines some of the knotty problems inherent in a quantitative approach which have not received adequate attention. The balance of this chapter is devoted to these problems.

At the outset the historian interested in quantitative analysis must grapple with the problem of making valid generalizations with historical data. Historical research is the interaction among the historian, his theories, and his data in a search for meaningful patterns of description and explanation of past events, ideas, actions, institutions, and processes. These patterns of description and explanation in fact are generalizations.[9] The problem for historians is not whether to generalize or not, but rather what constitutes an adequate generalization. This is a broad question, and our discussion focuses only on the most salient features of quantitative methods in historical analysis within the context of generalizations in history.

It should be noted that this emphasis raises some philosophical questions which we have chosen to keep in the background lest we become tangled in a web of epistemology. This does not mean that we consider such questions unimportant; on the contrary they require a more extensive treatment than can be offered here.

Quantitative Analysis and Measurement

At the lowest level of definition quantitative analysis means counting and comparing things which can be counted and compared. Since this requires the use of numbers, some historians claim that qualitative data expressed in words differ radically from quantitative data expressed in numbers. According to this view, there is a fundamental dichotomy between qualitative distinctions of kind and quantitative distinctions of amount, with the latter somehow being inferior. In this regard, it is worth noting that philosopher Abraham Kaplan argues quite persuasively that this dichotomy is more a matter of semantic confusion than a real distinction.[10] The real objection to using quantitative data does not seem to be numbers or counting, but rather the kind of measurement numbers imply. Presumably the assignment of numbers to phenomena according to specified rules renders qualitative historical data sterile.

Does the use of numbers require a kind of measurement that differs from that which historians typically use? All historians using documents measure, with or without numbers, when a research problem is defined as significant, information is selected as relevant, and comparisons are begun. Examination of the background of the gentry, or the ideology in American Revolutionary propaganda, or the connections between national character and economic abundance require some sort of measurement scale. Implicit use of a measurement scale does not alter the fact that differing characteristics of variables are specified. Unfortunately, implicit measurement usually is crude, uncontrolled, and subject to serious misunderstanding and error. Explicit rules governing the assignment of numbers to represent varying characteristics of certain properties elevates measurement to a higher level simply because it is an explicit objective, and standardized operation.[11] Once numbers have been assigned explicitly, they can be subjected to various modes of mathematical analysis. The patterns these modes of analysis disclose enable one to distinguish between different and similar phenomena within large data sets. The end result is a more accurate assessment of relationships which is essential to sound generalizations. Thus the rationale for the use of numbers in measurement is that in answering certain types of questions they enhance precision in assessing relationships and promote unambiguous communication.

Before using data already in numerical form or data to which numbers have been assigned according to certain rules, there are two questions which must be answered: what is measured? and what constitutes reliable and valid measurement? These two questions require us to consider briefly the philosophical theory of the meaning of words and concepts as they relate to research. Most historians use words and concepts as they find them in primary or secondary sources, with little appreciation of the philosophical questions involved. The philosophical theory of the meaning of words and concepts becomes relevant to historical research when the goal is to obtain reliable information which can be communicated clearly. By reliable information we mean results which may be replicated by other researchers, given the same research design and data.

At the outset let us distinguish between words and concepts. Many words function as names or labels which we assign to ideas, activities, and other phenomena. Thus, words help us to organize our perceptions of phenomena. Concepts, on the other hand, are rules governing the use of words. Since many words have more than one meaning, concepts specify the observations which identify the phenomena to which each usage applies. For example, one concept of revival refers to a new, emotional religious experience, while another refers to a deliberate restoration of an old pattern of activity.

Since concepts are rules for using words, they are neither true nor false, nor are they statements in themselves. However, concepts are used in statements against which evidence for or against can be brought to bear. Consider, for example, the use of the concept "charismatic" in a statement like "George Washington was a charismatic president." The accuracy of this statement can be tested by finding and evaluating documentary evidence to see if Washington's behavior and the popular response to him justify use of the concept. This theory of concepts has a direct bearing upon the question posed earlier about what is measured. When we talk about measuring something, we mean measuring directly or indirectly observable features of reality to which a concept refers. In other words we measure the degree to which empirical indicators of a given concept are present or absent. Thus, we measure the phenomena to which concepts refer, and the concepts determine the phenomena we look for.[12]

The measurement of phenomena to which concepts are referents involves the question of validity and reliability. Reliability means accuracy and validity means nonambiguity. Suppose that a measurement of the concept "political discontent" uses the Socialist vote in the 1912 election. The Socialist vote in this election is a *reliable* indicator of "political discontent" because the total number of Socialist votes cast will be nearly the same for each historian who tries to add them up. On the other hand, the Socialist vote is not a *valid* indicator of "political discontent" since many voters could have supported or rejected Socialist party candidates for reasons other than political discontent. In this instance our measurement of the phenomena to which "political discontent" refers is reliable but not valid.

Unreliable measurement is the result of selecting data which are incomplete for the particular concept employed. Invalid measurement stems from fuzzy thinking about the concept. Both reliable and valid measurement, therefore, require clearly formulated concepts and explicitly stated measurement techniques.

What criteria should be followed in formulating concepts which will yield valid and reliable measurement in historical analysis? First, each key concept should be divided into enough components to cover the phenomena to which the concept refers. These components may be derived logically from the concept or one component may be deduced from another. Regardless of the source, the components should be meaningful and agree logically. Second, reliable and observable indicators of each valid component must be found. If a reliable indicator(s) cannot be selected for a component, then it must be either deleted or used with the full awareness of inaccurate measurement. It is this test for correspondence between a concept or its components and specified reliable indicators which helps to impart objectivity to analysis. Finally, the assumptions built into a concept should be carefully examined and delineated.[13]

The formulation of concepts along these lines in historical analysis is clearly related to the use of numbers in measurement. Even if the empirical indicators of a concept are qualitative rather than quantitative, in most cases numbers can be assigned. However, this is not to say that because numbers can be assigned they should be used. In some instances the amount of time required to develop valid numerical indices may be excessive in terms of what they can reveal. In the final analysis the kinds of questions being asked and the quality of the data determine whether or not numerical measurement should be employed.

Quantitative Analysis and Statistics

Since quantitative analysis uses numerical data, some attention must be given at this point to statistics. Statistics can refer to numerical data, to measures derived from numerical data, and to techniques for extracting information from numerical data. As a general rule the context will make clear the intended meaning. Our consideration at this point will be limited to a brief exposition of several principles underlying those statistical techniques which relate to generalizations in history.

Statistics can be as esoteric and mathematical as one cares to make it—we plan to make it simple and straightforward. A useful distinction is that of description and prediction. Predictive statistics is of basic importance in experimental research but has limited application in historical research, especially in developing sound generalizations. On the other hand, descriptive statistics consists of techniques for describing numerical data in terms of central tendencies, variances, and relationships among variables. Since sound generalizations depend upon an accurate description of a body of data, the capability of descriptive statistics to summarize patterns implicit in a body of data lends itself to historical research. The various descriptive statistics techniques involve descriptions of either the distribution of one variable or relationships between two or more variables. The "average" or arithmetic mean, which yields a single value that summarizes a set of numbers, is a description of the distribution of the numbers. In addition to the average, a whole battery of techniques which measure various characteristics of a distribution can be called upon. The second category of descriptive statistics measures the relationships among variables. One such statistic is the coefficient of correlation between two or more variables. The mathematical concepts underlying the coefficient of correlation lead to a variety of techniques for measuring the degree of similarity and dissimilarity between sets of observations.

As we noted previously, descriptive statistics permit generalizations about specified characteristics of a body of data. Actually, there are two types of generalizations that can be made. The first is called a generalization about a population. In statistics, the word population does not refer to people, but includes the total class of whatever is observed as part of the study. A body of data can be called a population when it includes common measurements on a complete set of objects. For example, the presidential election returns from the state of Pennsylvania for the years 1824 through 1968 would constitute a population. The same information broken down into returns by county would make up a different population. The definition of a population is entirely arbitrary and is determined by the questions being asked. Regardless of the scope of a population, any generalization drawn *applies only to it*.

Sometimes it may not be practical to obtain numerical data for an entire population. In this circumstance a sample of the entire population is selected in such a manner as to represent accurately the population itself. One can then use the descriptive statistics about the sample to generalize about the population. In order to insure that these projected generalizations hold true for the population, the sample must be large enough and unbiased. An unbiased sample is one in which every element making up the population must have an equal mathematical chance to be included in the sample. A sample meeting this criterion is called a random sample.

Technically the only perfect random sample comes from numbering each unit of a population and then drawing a specified set of different numbers from a table of random numbers.[14] Unfortunately this procedure usually is too expensive and time-consuming for practical purposes. It can be approximated by systematically taking, say, every tenth name in a list of names—for example, a directory or a census manuscript. One minor drawback to this form of systematic sampling is that two relatives with the same surname probably will not appear together in the sample.

More elaborate procedures involve two or more steps. The first step may be to select "clusters" (for example, census tracts) at random, and then systematically take, say, every tenth unit in each cluster. Another first step might be to "stratify" the population into categories (Republicans, Democrats, and Independents) and then systematically sample from each category a number of units proportional to the relative size of the categories.

It is important to use cluster sampling carefully. A random sample of all Illinois voters cannot be achieved by selecting 20 counties at random and drawing 50 names from each. Chicagoans who constitute half the total population may never get represented. In this case it is wise to stratify first—that is, take 50 percent of the sample from Cook county and appropriate percentages from downstate rural and metropolitan counties. This insures that the sample more closely resembles the population.

The problem of sampling from incomplete or missing data has no simple solution. Perhaps the documentary data originally existed for a population, but has been repeatedly decimated by wars, fires, and carelessness. It may be plausible to assume that the ravages of time operated at random, and that a random sample from the incomplete data will adequately represent the original population. On the other hand, the original records may have been badly biased. Court records underrepresent honest men, while wills overrepresent old and wealthy men. The existence of bias can be checked by reference to another file of records known to be biased. For example, if colonial officeholders are known to have been richer than the average and to have constituted 10 percent of the adult males, then the names of the officeholders can be checked against the wills. If more than 10 percent of the wills represent officeholders, then the men who left wills were indeed richer than the average. The bias might not be so great as to invalidate the study, and sometimes it is possible to correct for the bias. One solution is to use a second file of data which will stratify the population into poor, middle, and rich, and then adjust a "stratified sample" drawn from the incomplete data so that it contains corresponding proportions of poor, middle, and rich.

The statistical properties of random samples have been the focus of a great deal of statistical theory.[15] The size of the population has little effect on the properties of a random sample drawn from it. Indeed, the accuracy of a random sample depends on the number of units (N) included in the sample. Suppose that the problem is to estimate what proportion (P) of a population possesses a certain characteristic. A sample of N units is drawn and a simple count shows that a certain number (K) of the sample units possess the characteristic. The best estimate of P therefore is $K/N = P^*$. Of course, P^* will rarely be exactly equal to P. The error can be quite high. However, the law of large numbers says that the value of P^* will usually be close to P, and in fact the larger the value of N the smaller the error will be. Table 1.1 shows the likelihood that errors of a specified size will occur for various values of N. The table can be interpreted simply. The entry for the row marked $N = 250$ and column 20% is ±4.0. The ±4.0 indicates that only 20 out of 100 different random samples of size 250 will have an error of more than 4 percentage points, while the other 80 samples will have an error of less than 4 points. The entry ±5.2 in the second column of the same row indicates that only 10 out of 100 different rar⁻ ᵈom samples of size 250 will have an error of more than 5.2 points either way. Hence, if $P^* = 60.0$ percent for a particular sample, it is possible to say that the chances are 80 in 100 that the true value of P lies between 56.0 percent and 64.0 percent; the chances are 90 in 100 that P lies somewhere between 54.8 percent and 65.2 percent; and the chances are 95 in 100 that P lies somewhere between 51.9 percent and 68.2 percent. Again, it must be emphasized that the sampling procedure must be random, and that the value of N refers to the size of the sample, not to the size of the population.[16]

How large should N be? The optimum size depends on the degree of accuracy desired and the expense involved in drawing a larger sample. Table 1.1 shows that when N is quadrupled, the maximum error shrinks by only one-half. Thus, an increase of N from 250 to 1000 (which involves a great increase in cost) only reduces the margin of uncertainty (at the 20 percent level) from 4 points to 2 points. On the other hand, a large sample may be necessary so that key segments of the population have adequate representation. Even when the sample size is large, say 1000, it often happens that an especially interesting subgroup is represented by only 50 or 100 units, and thus the possible range of error (even at the 20 percent level) is so large that the results are highly dubious.

TABLE 1.1 Level of Error in Sampling Proportions

SAMPLE SIZE N	20% (FOR 20 SAMPLES IN 100)	10% (FOR 10 SAMPLES IN 100)	5% (FOR 5 SAMPLES IN 100)	1% (FOR 1 SAMPLE IN 100)
2000	±1.4%	±1.8%	±2.2%	±2.9%
1000	2.0	2.6	3.1	4.4
500	2.9	3.7	4.4	5.8
250	4.0	5.2	6.2	8.2
100	6.4	8.2	10	13
50	9.1	12	14	18
25	13	16	20	26

A random sample may not accurately reflect the patterns in a given population. The proportions, means, and strength of association for two variables observed in a sample may differ widely from those in the population because certain atypical units are overrepresented in the sample. "Tests of significance" are statistical procedures which give the probability that a pattern in a random sample also exists in the population.[17] Usually the tests are used to indicate which observed sample patterns could be drawn from a population without these patterns once out of 10, 20, or 100 random samples of that population. Most statistical textbooks emphasize the use of tests of significance in analyzing experimental data rather than documentary data.

A crucial question for historians is when to use tests of significance. As a general rule tests of significance are inappropriate in historical research. Historians usually work with a population (a legislative body for example) or incomplete data from which a true random sample cannot be drawn. Tests of significance are valid only for random samples. Another stricture on the use of tests of significance is that they are quite sensitive to the size of the sample. When the N is small, a sample with a strong pattern could be drawn from a population without that pattern. For example, a correlation coefficient of +.8 could appear in a small sample, when in fact the true correlation in the population might be +.00. When N is larger, the probability is smaller that sample patterns will be grossly misleading. However, as the size of N increases, it is likely that even weak sample patterns (say $+.2$) can be judged significant. Hence, when N is large, tests of significance will nearly always show "statistically significant" results. Thus a test of significance can never prove that a pattern is significant in the sense that a historian should pay attention to it. It only means the improbability of selecting such a pattern at random. Only the scholar who understands his material can judge the intrinsic significance of patterns; tables in the back of statistical books are not equal to the challenge.

Quantitative Analysis and a Scientific Approach

Quantitative analysis requires the systematic study of the interrelationships of rigorously defined variables. Quantitative analysis, therefore, approximates a scientific approach to the extent that: (1) it rests upon a sound theoretical framework; (2) it utilizes models which make explicit the assumptions stemming from the theory; (3) it involves the development of testable hypotheses; and (4) it yields analytical results which other researchers should be able to replicate. These four elements comprise the basic dimensions of a sound research strategy. As a general rule historians have been derelict in developing sound research strategies, with embarrassing results to a profession so strongly committed to analyzing empirical data.[18] The use of a carefully drawn research strategy can enhance communication because other historians can comprehend exactly what has been attempted, what has been done, and how the conclusions can be used in their own work. Another advantage to formulating a research strategy is that of reducing the credibility gap in many published historical studies. By this we mean the literary advantages in avoiding explicit methodological considerations in polished historical narrative deprive readers of an adequate basis for assessing conclusions. While the conclusions may seem intuitively sound, their validity is determined by the procedures followed. A good example of intuitively plausible conclusions derived from inadequate research procedures is the Mowry-Hofstadter "status revolution" thesis of progressivism.[19]

A scientific approach in historical analysis calls for linking research with theory.[20] Theory here means a structured intellectual framework which is part of a larger body of knowledge. Traditionally, historians have paid scant attention to formalizing explicitly the intellectual framework with which they approach a problem. This reflects a naive but well meaning quest for objectivity as well as fuzzy thinking about the mental processes required for any analysis. Let it be clearly understood that history is not discovered in the sense that a historian approaches a question with a "tabula rasa," or that facts speak for themselves.[21] On the contrary, history is the product of certain assumptions and expectations about human behavior in general and a given set of phenomena in particular. For example, documentary evidence such as diaries or letters are usually considered superior sources because they often reveal declared intentions, goals, purposes, insights, and explanations of contemporaries. This evaluation rests upon a theory of rational human behavior which postulates human behavior as a conscious goal-seeking process. Consequently, the historian may accept the document pretty much at face value if it satisfies the canons of internal and external criticism. Another general theory would rely far less on conscious motivation to explain human behavior. Instead, there are certain biological, psychological, and environmental factors which motivate all people. Thus, the real explanation is more likely to be an unconscious motivation which is glossed over in diaries and letters. This theory of "unconscious motivation" as being the "real" cause of human behavior provided the underpinning of Charles A. Beard's economic interpretation of the origins of the Constitution. The true explanation of the founding father's actions, he believed, was to be found in economic holdings rather than in polemics over ratification. For this reason Beard's data consisted of an "economic biography" of each of the 55 delegates who attended the Philadelphia Convention.[22]

Implicit theories of probable human behavior frequently govern the evaluation of conflicting evidence from the past. Vernon K. Dibble has made explicit the probability statements which might form the basis for evaluating two differing accounts of Patrick Henry's speech in the House of Burgesses on May 30, 1765.[23] According to Thomas Jefferson some 40 years after the event, Henry closed his speech with "If this be treason, make the most of it." A French traveller who was present recorded in his diary that Henry responded to the charge of treason by declaring his loyalty to the King even "at the Expense of the last Drop of his blood." Dibble suggests that in evaluating the two accounts the historian might invoke certain probability statements about human behavior.

> . . . disinterested testimony is likely to be more accurate than politically charged testimony, testimony to words spoken in one's native language is likely to be more accurate than words spoken in another language, unprompted testimony is likely to be more accurate than prompted testimony (Jefferson was asked by William Wirt if the legendary version of the speech were not true), and testimony recorded before many versions of the event in question have been heard is likely to be more accurate than testimony recorded after the witness has heard a variety of versions.[24]

Our concern here is not which of the two accounts is more accurate, but rather to emphasize that in assessing evidence historians continually utilize such implicit theories of probable human behavior.

As a third example of the use of implicit theory in historical analysis, consider the presuppositions one must make, say, in studying a social institution or a political movement. One way to approach a study of either topic requires the assumption of the regularity of occurrence of certain expectations and actions among a group of individuals. In other words, one assumes that the participants shared mutual perceptions of expected behavior and responded in the expected manner. This approach, which we may call collectivism, derives from a view of social structure "composed of interlocking roles performed by individuals with reciprocal expectations and resulting in mutually adjusted behavior."[25] Another basic approach begins with the idea that what appears to be group behavior is really the aggregating of similar acts of single individuals without any conscious planning.[26] Aggregative group behavior, therefore, is not the conscious product of the actors, but of someone who subsequently groups together what appears to him to be similar ideas and actions. A third approach, which Robert Berkhofer calls "conjunctive," stands between collective and group behavior.[27] Aggregative phenomena such as business cycles may result in collective behavior. Likewise, collective behavior may yield unanticipated results which are aggregative. Thus the conjunctive approach combines the viewpoints of the actor and the observer.

At this point one might object that our discussion of theory is irrelevant since what the historian actually does is to rely upon generally accepted experience called common sense. Since common sense accurately reflects the normal experience of everyday life, the historian need not concern himself with theory. Common sense is equal to, or superior to, theory. This is a fallacious notion. To be sure, many times common sense assumptions and statements are correct. Yet, common sense experience tends to oversimplify through failure to consider all factors. In addition, the vague or assumed definitions in many common sense statements confuse rather than enlighten. In short, common sense really is low level theory which has not been systematically organized and formulated.[28]

What are the advantages of using explicit theory in historical analysis? First, it forces one to examine critically the presuppositions underlying a study. This helps to identify the extent to which a researcher's social interests and own point of view bias his work. Second, it ensures that proper attention will be given to the way theory affects the formulation of research problems. In historical analysis, as in other disciplines, the development of a research problem, which includes the questions asked as well as how they are to be answered, is the product of the researcher's theory—explicit or implicit. Explicit theory helps a researcher to frame a significant research problem in terms of a set of unambiguous questions asking what relation holds between two or more specified subjects or items. This guides historians to use hypotheses that refer to observable, measurable events. In other words, theory generates hypotheses which in turn determine the appropriate data required to test adequately the hypotheses. Third, explicit theory provides a logical and relevant connection with a larger interpretative framework which can be applied, or is applied, to other problems. This directs historians to recognize the importance of the interconnectedness between guiding principles in the social sciences.

In addition to linking research with theory, a scientific approach in historical analysis utilizes an explicitly formulated model.[29] A model includes what the general theory suggests to be the important subjects and conditions while ignoring all the unimportant subjects and conditions. It embodies the research problem questions by defining what relationships should exist among subjects and under what conditions. Much confusion and misunderstanding can be avoided by recognizing that a model is an analogy which is similar in some, but not all, respects to the subjects and conditions being studied. Like any analogy, therefore, it is a simplification of complex relationships and is not isomorphic to the real world. Nevertheless, a carefully formulated model is extremely useful in historical analysis. It delineates the implications of a general theory when it is applied to a concrete historical situation. It helps to clarify relationships among subjects and demonstrates that these relationships are logically consistent. And as we show later, the use of such a model generates testable hypotheses.

While models can be differentiated in several ways, the most useful distinction for historical analysis probably is verbal and mathematical. By verbal we mean that the model consists of one or more logically consistent statements which specify relationships among certain phenomena. Richard Hofstadter uses an elementary verbal model in his study of Progressivism which stipulates that certain middle class persons behaved like reformers under conditions of status decline. An example of a more involved verbal model is Anthony Down's model of the democratic state, which he derived from economic theory.[30] The model states that: (1) when a person votes he seeks to maximize his personal utilities; (2) each political party seeks to win and maintain office; (3) both parties and voters pursue these goals rationally.

The main difference between a verbal model and a mathematical model is that in the latter the basic relationships among subjects are expressed mathematically by, say, a correlation coefficient, a regression analysis equation, or a set of complicated equations. A mathematical model of Hofstadter's status-anxiety verbal model of progressivism might be a correlation coefficient measuring the association among "profile scores" of a set of middle class political reformers around the turn of the century. A much better example of a mathematical model in historical analysis is found in the study of the profitability of slave operations in the antebellum South by Alfred H. Conrad and John R. Meyer.[31] The model required computation of the capital-value formula, $Y = X_t/(1 + r)^t$. In the formula Y = present value, X_t = realized net return t years hence, and r = interest rate. To compute this formula the authors had to determine the longevity of slaves, the capital investment in slaves (that is, initial cost of slaves, land, and equipment), the interest rate, and the annual returns of slave production in the field, and procreation after deduction of costs of slave maintenance. An example of a very elaborate mathematical model is the one Hayward R. Alker, Jr.[32] formulated for Daniel Lerner's verbal model of modernization.[33] Lerner's model expresses a cause-effect relationship in which increased urbanization raises the literacy rate, literacy then enlarges media development, which in turn boosts political participation. Mathematically Alker represented the model with the following equations:

$$
\begin{aligned}
X_1 &= U_1 \\
a_{21} + X_1 + X_2 &= U_2 \\
a_{32} + X_2 + X_3 &= U_3 \\
a_{43} + X_3 + X_4 &= U_4 \\
\Sigma U_1 U_2 = \Sigma U_1 U_3 = \Sigma U_1 U_4 &= \Sigma U_2 U_3 \\
&= \Sigma U_2 U_4 = \Sigma U_3 U_4 = 0
\end{aligned}
$$

X_1 denotes urbanization, X_2 literacy, X_3 media development, and X_4 political participation. U_1, U_2, U_3, and U_4 are uncorrelated residual causes while a_{21}, a_{32}, and a_{43} are dependence coefficients.[34]

There will probably be few times that historians have need for a mathematical model as sophisticated as Alker's. As a general rule, a simple and straightforward mathematical model will suffice. The chief reason we introduced Alker's model was to illustrate how the rigor and precision of a mathematical model can expose a deficiency in a verbal model. Alker's calculations using the above equations did not produce the anticipated results, thus indicating a weakness in Lerner's model. A revision of the verbal model with appropriate alterations in the corresponding equations so that increased urbanization caused the development of both literacy and media, with the former playing the "key role" in increasing political participation, yielded more acceptable results.[35]

The third element of a scientific approach is the development of testable hypotheses. Broadly speaking, a hypothesis is a tentative generalization about a problem under investigation, the tenability of which is determined by compatibility with empirical data. Certainly, this should not be something new to historians. We use hypotheses implicitly each time we research a problem in history. However, a scientific approach requires the explicit formulation of testable hypotheses. Testable hypotheses specify what the anticipated relations are, spell out working definitions of key concepts, and define the empirical indicators of the concepts. Hypotheses in historical research not satisfying these criteria are not amenable to a scientific approach.

Some historians may object to explicit hypotheses, claiming that they bias investigation and confine research to such a narrow and limited scope that more important things are missed. The notion that hypotheses prevent objectivity and blind the investigator to "big" issues and questions reflects a serious misunderstanding of the impact of unarticulated assumptions and hypotheses upon research. Biased investigation and missed "big" issues and questions are far more likely to occur when the research is guided by vague, obscure, and unexamined ideas and questions than when it is guided by carefully expressed hypotheses. Lee Benson's seminal essay, "Research Problems in American Political Historiography," makes this abundantly clear.[36] An excellent example of how carefully thought out hypotheses can strengthen historical analysis is the Conrad and Meyer study of the profitability of slavery in the antebellum South. The authors hypothesized that "slavery was an efficient, maintainable form of economic organization."[37] They defined a slave economy in terms of two production functions. "One function relates inputs of Negro slaves (and the material required to maintain the slaves) to the production of southern staple crops, especially cotton. The second function describes the production of the intermediate good, slave labor-slave breeding. . . ."[38] Viewed from the outside as an efficient economic organization, slavery had to yield a return comparable to that being earned on other capital assets at the time. Internally slavery could be efficient only if "those regions best suited to the production of cotton (and the other important staples) specialized in agricultural production, while the less productive land continued to produce slaves, exporting the increase to the staple-crop areas."[39] Conrad and Meyer pointed out the implications of their hypotheses and definitions in terms of

> . . . putting appropriate values on the variables in the production functions and computing any present value over cost created by the stream of income over the lifetime of the slave. These returns must, of course, be, shown to be at least equal to those earnable elsewhere in the American economy at the time. It is further neces-

sary to show that appropriate slave markets existed to make regional specialization in slave-breeding possible and that slavery did not necessarily imply the disappearance or misallocation of capital. Evidence on the ability of the slave force to maintain itself numerically will be had as a corollary result. To accomplish all these assessments, it is necessary to obtain data on slave prices and cotton prices, the average output of male field hands and field wenches, the life-expectancy of Negroes born in slavery, the cost of maintaining slaves during infancy and other nonproductive periods, and finally, the net reproduction rate and the demographic composition of the slave population in the breeding and using areas.[40]

The quotations and excerpts illustrate at least six contributions testable hypotheses can make to historical analysis. These contributions can be summarized as follows: (1) testable hypotheses are a bridge between theory, models, and the actual research activity; (2) testable hypotheses define the direction, focus, and goals of a study; (3) testable hypotheses provide criteria by which to select meaningful data; (4) testable hypotheses help to clarify the full dimensions of a problem by utilizing evidence which might be contrary; (5) testable hypotheses develop implications which can lead to rejection; and (6) testable hypotheses offer a general framework for drawing credible conclusions.

There are few ready made testable hypotheses at hand for use. As a general rule testable hypotheses are drawn together from several sources. One source, as we suggested earlier, is the conscious articulation of the theory and model underlying the investigation. A hunch about relations among phenomena is another source. Reading the literature in a field frequently will reveal statements and assertions which really are untested generalizations. A creative mind which can see new lines of inquiry in old paths of research will come up with interesting testable hypotheses, as did Conrad and Meyer. More than likely the translation of inchoate ideas into testable hypotheses will be the product of the interaction among these sources and others.

How does one actually test a hypothesis? Suppose that we have a testable hypothesis that in the 1840s in Mississippi most Whigs were wealthy while most Democrats were poor or only moderately well off. (Of course we have already defined Whigs, Democrats, and wealth and selected the appropriate data.) In order to support this hypothesis we must reject its opposite, that is, the null hypothesis that there is little difference in the number of men of high, middle, and low income in the two parties. If the hypothesis were that there is little difference in the number of men of high, middle, and low income in the two parties, the null hypothesis would be the original hypothesis. A hypothesis is either accepted or not on the basis of a decision about its logical alterna-

TABLE 1.2   Checklist for Errors in Research Design

Error Types: *A* Routine mistakes
 B Conceptual mistakes
 C Statistical mistakes
   alpha—reject a true hypothesis
   beta—accept a false hypothesis

STAGE OF RESEARCH PROCESS	ERROR TYPE	ERROR
1. Planning the project	*A*	Overlooking relevant scholarship.
	B	Inadequate attention to theory, models, and hypotheses; asking unimportant or misleading questions; selecting wrong variables or poor indices; overextending scope of project relative to resources available.
2. Locating the historical data	*A,B*	Overlooking valuable data; underestimating error in source.
3. Sampling procedure	*A*	Sample not random or wrong population used.
	B	Sample too small for analysis or too large for budget; not recording the right data.
4. Transcribing data, coding, and machine processing	*A*	Mistakes in transcription; cluttered IBM cards; destroying the data source.
	B	Awkward codes; vague or censored data problems; bad program.
5. Computing statistics	*A*	Arithmetic mistakes, misreading tables and formulae.
	B	Using wrong tests.
6. Analysis of statistics	*B*	Overlooking controls; misinterpreting a test.
	C alpha	Reject true hypothesis
	C beta	Accept false hypothesis
7. Conclusions drawn	*B,C*	Overinterpretation, reading too much into the results; overgeneralization; failing to end successful study; failing to modify an unsuccessful study; failing to continue a significant study.
8. Publication of results	*A*	Mistakes in transcription.
	B	Not including enough techniques or data to permit others to replicate or validate the research; suppressing counter-evidence; exaggerated claims.

tive—a null hypothesis. Rejection of the null hypothesis does not prove the original hypothesis in the sense of disproving every other potential null hypothesis which might be tested in this connection. The chief function of the null hypothesis is that of providing a reasonal basis for establishing the credibility of the original hypothesis.

What does one do when the null hypothesis is confirmed rather than the original hypothesis? This may be a matter of concern to those historians who think a negative finding is less desirable than a positive one. In our opinion the elimination of a potential hypothesis is in fact a significant accomplishment. However, before accepting the negative findings as final, each stage of the research design should be carefully checked for errors. Table 1.2 is a useful checklist to follow in trying to locate an error. If an error is located, the appropriate adjustments should be made and the hypothesis tested again. If no error is located, the negative finding can be converted into a positive one simply by making it the hypothesis to be tested. Since hypotheses are only tentative statements, they should not be considered sacrosanct.

We began this section on quantitative analysis and scientific approach with an emphasis on research design. The three levels of this approach we have discussed embody the major themes of a research design. A research design should be written out and incorporated into the final published study. Readers are entitled to know not only the conclusions but also how the conclusions were reached.

FOOTNOTES

1. Frederick Jackson Turner, *The Significance of Sections in American History* (New York: 1932) pp. 7, 20; emphasis added.

2. Avery Craven, "Frederick Jackson Turner," in William T. Hutchinson, ed., *The Marcus W. Jernegan Essays in American Historiography* (Chicago: 1937) p. 265.

3. Quoted from class notes in Joseph Schafer, "The Microscopic Method Applied to History," *Minnesota History Bulletin,* 4 (1921) p. 19; this is one of the few methodological essays published by the Turnerians.

4. Turner, *Significance of Sections,* p. 45.

5. Henry Cabot Lodge, "The Distribution of Ability in the United States," *Century Magazine,* 42 (1891), pp. 687–94. Lodge in turn relied on an inferior study of British elites by Arthur Conan Doyle.

6. ORIN G. LIBBY, "A Plea for the Study of Votes in Congress," *American Historical Association Annual Report for 1896* (Washington: 1897) v. 1, pp. 321–34.

7. WILLIAM F. DICKINSON, "The Personnel of the Wisconsin Legislature for the Years 1850, 1860, and 1870," (B. L. thesis, University of Wisconsin Library, 1901). The extant copy bears the approving signatures of Libby and Turner, but lacks an important summary chart.

8. For details see Richard Jensen, "History and the Political Scientist," and "American Election Analysis: A Case History of Methodological Innovation and Diffusion," in Seymour Martin Lipset, ed., *Politics and the Social Sciences* (New York: 1969).

9. For a general treatment see *Generalization in the Writing of History,* ed. Louis Gottschalk (Chicago: 1963), especially the essay by William O. Aydelotte, "Notes on Historical Generalization."

10. ABRAHAM KAPLAN, *The Conduct of Inquiry* (San Francisco: 1964), p. 207.

11. These rules of assignment are subsumed under scales of measurement called dichotomous, categorical, ranked, and interval. They are treated in detail in Chapter 2.

12. In a general sense the measurement of empirical indicators is a "behavioral" approach—that is, it depends on what actually happened, not on a normative or counterfactual analysis of what should or might have happened. In this text we avoid the term "behavioral" because its use in some social sciences, especially social psychology and political science, is closely tied to the theory that people's attitudes are the basic unit of social analysis. Behavioralism in those disciplines was, historically, closely linked to the use of statistics, and a confusion has persisted to the effect that quantification requires a "behavioral" theoretical foundation. See James Charlesworth, ed., *Contemporary Political Analysis* (New York: 1968); and generally Robert Berkhofer, *A Behavioral Approach to Historical Analysis* (New York: 1969).

13. This aspect of concept formulation is discussed in greater detail later in this chapter.

14. Most books on statistics contain a table of four digit random numbers. The use of this table is quite simple. Suppose that there are 450 units in the population and that the sample will contain 100 units. Begin anywhere in the table and examine the first three digits in each column. Moving down each column select any set of three digits from 001 to 450 until 100 different sets have been chosen. (Digit sets from 451 to 999, of course, will be ignored.) Each set of three digits corresponds to one of the 450 units in the population and the 100 units are a random sample of the total population.

15. For sage advice see V. O. Key, *A Primer of Statistics for Political Scientists* (New York, 1954), ch. 6, and the *International Encyclopedia of the Social Sciences* (1968) article on "Sample Surveys: Nonprobability Samples." Full technical discussions can be found in William E. Deming, *Some Theory of Sampling* (New York: 1950) and William G. Cochran, *Sampling Techniques* (New York: 1968).

16. The formula for the sampling error in estimating interval variables is more complex. Suppose the mean value of X for a random sample of size N is $\bar{X}$, and the true target population has mean M. Then in 80 out of 100 different random samples, M lies somewhere between $\bar{X} - 1.28$ (st dev X)$/N$ and $\bar{X} + 1.28$ (st dev X)$/N$ where st dev X is the standard deviation of X for the sample (and can be computed by the methods discussed in Chapter 3), and 1.28 is a coefficient related to the 20% probability level. The coefficients for the 10%, 5%, and 1% probability levels are 1.64, 1.96, and 2.58, respectively. Historians who fear that their sample size N is too small to make reliable estimates of the target mean should consult a standard statistics textbook for details on the testing procedures sketched here. For example, Frederick Croxton, Dudley Cowden, and Sidney Klein, *Applied General Statistics* (3rd ed, Englewood Cliffs, 1967), ch. 24 and 25 for step-by-step procedures.

17. For a readable and systematic discussion of the statistical problems involved, see William Hays, *Statistics for Psychologists* (New York: 1963), ch. 9 and 10. A lively discussion of the general problems of research problems involved appears in Travis Hirschi and Hanan Selvin, *Delinquency Research* (New York: 1967), especially ch. 13. The entire book repays close reading by historians, who should be able to transcend its narrow focus on sociological problems. For a theoretical treatment see John Galtung, *Theory and Methods of Social Research* (New York: 1967), 358–389.

18. An excellent example of the development of a sound research strategy is Alfred H. Conrad and John R. Meyer, *The Economics of Slavery* (Chicago: 1964).

19. George W. Mowry, *The California Progressives* (Berkley: 1951); Richard Hofstadter, *The Age of Reform* (New York: 1955). For criticisms of the status-revolution thesis see the following: Jack Tager, "Progressives, Conservatives, and the Theory of the Status Revolution," *Mid-America*, 48 (1966), pp. 162–175; William T. Kerr, "The Progressives of Washington, 1910–1912," *Pacific Northwestern Quarterly*, 54 (1964), pp. 16–27; Richard Sherman, "Status Revolution and Massachusetts Progressive Leadership," *Political Science Quarterly*, 78 (1963), pp. 59–65; Robert Skotheim, "A Note on Historical Method: David Donald's 'Toward a Reconsideration of Abolitionists,' " *Journal of Southern History*, 25 (1959), pp. 356–65.

20. The best available work on this subject is the previously mentioned book by Robert Berkhofer, *A Behavioral Approach to Historical Analysis*. This book fills a great void in historical methodology. The authors benefited greatly from Berkhofer's perceptive discussion of a number of topics crucial in historical analysis.

21. Our position here is not to be identified with a "Construction Theory of History" as developed by Jack W. Meiland in *Scepticism and Historical Knowledge* (New York: 1965).

22. For very illuminating treatments of Beard's approach see Lee Benson, *Turner and Beard: American Historical Writing Reconsidered* (New York: 1960); and William Appleman Williams, "A Note on Charles Austin Beard's Search for a General Theory of Causation," *American Historical Review*, 62 (1956), pp. 59–80.

23. DIBBLE, "Four Types of Inference from Documents to Events," *History and Theory*, 3 (1963), pp. 203-21.

24. DIBBLE, "Four Types of Inference," p. 205.

25. BERKHOFER, *A Behavioral Approach*, p. 81.

26. Notice how this relates to our previous discussion of unconscious motivation.

27. BERKHOFER, *A Behavioral Approach*, p. 77.

28. FOLKE DOVRING, *History As A Social Science* (The Hague: 1960). Also see Fred Kerlinger, *Foundations of Behavioral Research* (New York: 1964), pp. 4-6, for a succinct discussion of science and common sense.

29. BERKHOFER, *A Behavioral Approach*, pp. 169-88.

30. DOWN, *An Economic Theory of Democracy* (New York: 1957).

31. CONRAD and MEYER, *The Economics of Slavery*, pp. 43-84.

32. "Causal Inference and Political Analysis," in *Mathematical Applications in Political Science*, II, Joseph L. Bernd, ed., (Dallas: 1966), pp. 19-25.

33. *The Passing of Traditional Society* (New York: 1958), p. 56.

34. For an explanation of the equations see Alker, "Causal Inference."

35. ALKER, "Causal Inference."

36. In Mira Komarovsky, ed., *Common Frontiers in the Social Sciences* (Glencoe: 1957), pp. 133–182.

37. CONRAD and MEYER, *The Economics of Slavery*, p. 45.

38. CONRAD and MEYER, p. 45.

39. CONRAD and MEYER, pp. 45–46.

40. CONRAD and MEYER, p. 47.

ELEMENTARY DESCRIPTIVE STATISTICS

CHAPTER TWO

Before the standard techniques of data manipulation and analysis can be applied, the researcher must consider the validity and reliability of the numbers he is using. Validity refers to the relationship between the phenomenon being studied and the quantitative information used to describe it. The primary questions regarding validity are: can the patterns in the historical events be summarized in quantitative terms adequate to the research design, and if so what kind of measurement is involved in summarizing those patterns? The reliability of a set of data involves the errors and biases that have entered into the original measurements, and the biases that may result from imperfect preservation of the records, or from the sampling procedure used to reduce excessively large data files to manageable form. The general problems of validity and reliability were discussed in the last chapter. This chapter considers the more technical aspects of measurement and elementary pattern-searching techniques.

Measurement and Error

Valid data ready for analysis always appear as the measured attributes of a group of subjects taken with respect to precise dates and geographical locations. The "subjects" can be individual persons, institutions (like governments, firms, religious bodies, or political parties), events (wars, depressions, elections, roll calls), or groups of people or geographical units characterized by a common attribute (all Catholic men, the participants in major riots, the cotton-belt counties of a region). Actual words or phrases can become subjects if a content analysis research design provides coding procedures that count cases of explicit word usage or implicit rhetorical nuances in the language of written or spoken documents. The dates and geographical origin of the data sometimes pose technical difficulties, which must be resolved by the application of traditional historiographic criticism.

The "measured attributes," usually called "variables," include a surprisingly wide range of characteristics. Almost any attribute that a scholar can objectively assign to specific subjects is amenable to some form of quantitative measurement—indeed, any attribute that can be adjudged present or absent, in greater or lesser degree, can be quantified. Of course, whether anything is to be gained by reducing the complexity of intricate patterns by subjecting them to precise measurement is a basic decision that must be made when designing a research strategy. In general, the greater the amount of data, the greater is the need for quantification; conversely, the greater the complexity of the data, the more difficult is the establishment of definitions precise enough to allow valid measurement.

The distinction between "subjects" and "attributes" may appear scholastic, but it does facilitate the design of proper research strategies, especially when the choice must be made between different statistical methods. Attributes have variability (present-absent; high-middle-low), while subjects (the Catholic Church) do not. In fact, subjects cannot be quantified at all, only their attributes. Thus a group of men has no measurement per se, but the size of the group and its average age can be quantified. In data processing the subject is the carrier of the information to be analyzed—sometimes in a literal sense, for IBM cards are used to carry information about the quantified attributes of subjects, one subject per card. And in the final results all the conclusions must be expressed in terms of relationships among subjects, for they are the focus of all historical research.

Attributes can be quantified in terms of one or more of five levels of measurement: dichotomous, nominal categorical, ordered categorical, ranked, and interval. Different statistical methods are appropriate to each level of measurement.[1]

Dichotomous attributes are those which a subject either possesses or lacks, such as "Catholic" (vs "noncatholic"), "beheaded by the terror" (vs "not-beheaded"), "at war" (vs "not at war"), "gaining population" (vs "not gaining population"), or "voting Republican" (vs "not voting

Republican"). Dichotomous measurement gains convenience at the loss of the complexity of human events.

An attribute falls in the categorical levels if a rule is defined that places each subject into one, and only one of three or more possible classes. If there is a natural order to the classes, the level is ordered; if the sequence of classes makes no difference, the level is nominal. Thus the attribute "religion" becomes a categorical variable when a procedure is set up that classes each subject as either "Catholic," "Jewish," "Protestant," or "Moslem." Since there is no intrinsic order to these four categories, the variable is nominal. Wealth may become a categorical variable when a procedure is devised to class each subject as either "poor," "middling," or "rich." This time there is a natural order to the classes, for the sequence "middling," "poor," "rich" is illogical; hence the variable is ordered. Similarly a place can be classified as a hamlet, town, city, or metropolis by a suitable rule, or the impact of a war can be rated on scale from 1 (practically bloodless) to 10 (nuclear holocaust), like earthquakes, and the natural ordering indicates the existence of an ordered categorical variable. In roll call analysis, the actual votes are coded as nominal variables ("yea," "nay," "absent," "paired against," "not a member," and so on), while the Guttman scale positions that may be derived are ordered categories.

Everyone trying to establish a satisfactory rule for assigning subjects to categories encounters difficulties of classification. The problem is especially acute when the data are rich in connotations and the historian feels a natural urge to create special categories all the time to handle cases he knows to be unusual. No universal guidelines exist, but a few rules of thumb are useful. The classes should be precisely defined in terms of objectively determined criteria, the classes should not overlap, and there should be as few different categories as possible (to simplify the statistics). A rule that creates categories with about the same number of subjects in each is slightly preferable to one that has a lopsided distribution, or which assigns more subjects to a catchall "other" category. Note that "other" categories are usually too vague to be amenable to analysis.

Ranked variables are attributes that permit *all* the subjects to be arranged in ascending or descending order. In a sense they are equivalent to ordered categorical variables that have only one or two subjects in each category. Some attributes, such as prestige, skill, power, health, sanitary condition, beauty, popularity, seniority, and repulsiveness naturally form rankings in a group of subjects. Historians generally find ranked variables most useful as surrogates for interval variables when the latter are unknown or are too cumbersome for rapid statistical analysis. For example the urban historian may want to study the relationship between neighborhood wealth and voting behavior, but be stymied for lack of data about wealth. He may decide to rank the wards in order of infant death rate, on the basis of other evidence, and assume that this ranking is similar to the ranking of wards by wealth. Each ranked subject should be coded with an ordinal number first, second, third . . . last (written 1, 2, 3, . . . Nth), with first place assigned to either the highest or the lowest subject. In case of ties, each subject should be assigned an average rating; thus, if two subjects are tied for fourth and fifth place, assign each the rank 4.5, and if three subjects are tied for eighth, ninth, and tenth place, assign each the rank 9. If most of the subjects are tied two-way or three-way, the ranking collapses into the ordered category level of measurement.

Interval variables are the kind usually considered "measurement" by people unfamiliar with the other four levels. Interval variables always take precise numerical form—1,500,000 people, $1500 wealth, 75¢/bushel, 6% interest, one shilling per pound tax rate, 640 acres, 10 years duration, 16 vote majority, 2 inhabitants per square mile, or 500 kilometers of railroad track. Interval variables are derived from *counts* of subjects, from *percentages*, from *rates*, from *ratios*, and from *transformations* and *coefficients* whose values are estimated from a statistical or mathematical model. The different types of derivation are analytically similar and should present no particular difficulties.

Specious accuracy typically appears in published interval data (for example, "43,614,313" immigrants came to the United States from 1820 through 1966; "6,730,935" Englishmen were eligible to vote in 1900), and must be taken with a grain of salt and a dose of rounding. "Economics," Norbert Wiener once observed, "is a one or two digit science." History is hardly more accurate—nor need it be. The historian with an excessive amount of zeal and patience can calculate a man's age to the nearest hour, but the nearest year will always be adequate for research. Interval data should be rounded after the first three exact digits for purposes of statistical analysis. In published analytic tables even three digits can be cumbersome, and may usually be replaced by two digits. Rounding to two digits is a healthy experience, for it constantly reminds the researcher of the reliability of his data.

Usually the basic counts underlying interval variables were made by clerks or careful observers, not by the scholar who is asking his own set of questions. The sources of bias or error in counting must be evaluated to gauge the reliability of the data. A few rules of thumb may speed up the reliability check.

1.   *Tabulation and Typographical Errors.* Always scan the source for suspiciously large, small, or identical entries. Did the tabulating agency have adequate resources for careful work? Note that add-

ing machines and Hollerith (IBM) cards were first used by governmental statistical agencies in the late nineteenth and early twentieth century.

2. *Underreporting.* Check for missing entries. Look for strings of zeroes that indicate haphazard guesses ("500,000" population). Are rates in contiguous geographical areas or years comparable, or are there inexplicable differences? Are there erratic jumps or plunges in time series that probably should be smooth? What types of respondents may have failed to return a report, or have been overlooked? Was there a procedure to follow up missing returns, or were old data quietly used when fresh information was missing?

3. *Questionnaire Bias.* Were the questions posed in person by a trained enumerator or interviewer, or by political hacks or loafers who may have invented interviews. In population studies, were neighbors sometimes called upon to provide information they probably did not know? How difficult was it for a confused, hurried, or illiterate person to interpret the questions precisely and answer them accurately? (Many literate people today inaccurately report their income, schooling, and age! One gauge of the competence of the respondents in a census is the proportion who report their age in multiples of five or ten years.) Was there a penalty for nonresponse, or a threatened burden levied on the truth in the form of taxes or conscription?

4. *Falsification and Fabrication.* Did inducements exist for the exaggeration, falsification, or fabrication of individual or aggregate returns? Were canvassers paid by the number of responses they collected? Was political patronage involved? Did charges of widespread corruption attend the survey? Were the supervisors indifferent to the trends that might appear, or did their careers depend on favorable results? Did the sponsors and critics put great reliance on the results?

5. *Changes in Survey Design.* Was the geographical scope of the survey the same from year to year? Were the questions phrased the same way? Did the definitions of basic units change? (In 1900 the U.S. Census changed its definition of "urban" from 8000 population to only 2500.) Can unexpected jumps in time series be explained by changes in definition, or were real factors involved?

Error is inevitable in quantitative data—indeed in all data. It is true that a large sample is more reliable than a smaller one, if both are chosen by the same methods; but this does not imply that aggregate data for large units are more reliable than the same data for small units. The reliability of all data primarily depends not on quantity, but on the quality of collection.

Errors in the collection of data can be typed as "random" or "systematic." Random error occurs when the data collectors erred sometimes one way, and sometimes the other. If the number of observations is fairly large (more than 100 or so), and the returns are added together to give aggregate data, then random error tends to cancel out and poses no serious problem. Note that the same is not true for random errors in tabulations. A copyist or keypuncher is as likely to write 8,567,000 for 3,567,000 as he is to write it 3,568,000. The only remedy for errors of tabulation and transcription is discovery and correction. Systematic source error, which tends to add up instead of cancel out, results from the routine underreporting or exaggeration of the true values. Even good national censuses taken under scientific auspices miss 2 or 3 percent of the population by systematically undercounting the poor, transient, vagrant, or inaccessible back country inhabitants. The United States Census has always missed hundreds of thousands, even millions of Negroes, but it seldom overlooks higher status, wealthier, or more visible whites.[2]

Fortunately the presence of systematic error does not always invalidate conclusions drawn from data. If the systematic bias is the same for all areas, or for all of a time series, then the patterns of relationship among areas or years will not be seriously affected by the bias. For example, if the census overlooks 20 percent of the Negroes in each state, and also underreports the average income of each state by 10 percent, the correlation between these two faulty data sets will nevertheless be exactly the same as the correlation that would have resulted from absolutely accurate data. However if one bias is stronger in some areas, the correlation that results may be too high or too low. If random error seriously affects two data sets, the correlation coefficient between them will be lower (almost never higher) than the true correlation. Thus the historian who is aware of the presence of a high level of random error in a data set can be confident when he finds a fairly high correlation that the true correlation is even higher. In any case it is important to make a rough evaluation of the types of error present, and estimate which way the error will bias the conclusions to be drawn.

Percentages and The Slide Rule

A large table crammed with thousands of digits may contain a great deal of valuable information, but it will not convey that information to the researcher who is unable to process the numbers so as to highlight the important patterns. A variety of simple techniques are widely used to bring out the patterns. Graphs can display distributions of magnitude; pie diagrams can reveal the relative divisions for categorical variables; charts can show the structure of time series; statistical maps can reveal areal distributions. The simplest and most important technique for dichotomous and categorical data is the conversion of the data from raw numbers to percentages.

The calculation of percentages is an easy matter on a slide rule. A ten-inch plastic rule containing the most often used scales (A, B, C, CI, D, K, L, S, T) can be purchased for under three dollars, and will quickly become a valued companion through forests of numbers. Every new slide rule comes with a short guide that shows how to read the scales and multiply, divide, square, take square roots and reciprocals, and perform various less relevant operations. As a mechanical instrument the slide rule can perform two functions: comparison of different scales, and addition or subtraction of lengths. The slide rule cannot, however, indicate the location of decimal points, so these must be determined separately.

Some operations can be performed merely by moving the hairline or cursor, a sliding transparent device that helps locate the number on one scale that is directly above a number on a second scale. To square a number, locate it on the D scale and read its square directly on the A scale. To find the square root of a number, count the number of digits and zeroes it has to the left of the decimal point (or, if less than 1.0, the number of zeroes before the first nonzero digit). If the count is odd (for example, 4, 400.00, 40,000, or .04) the number must be found on the *left* half of the A scale. The correct square root (2, 20.0, 200, or 0.2) appears directly below on the D scale. If the count is even (for example, 49.0, 4900, or .49), the number must be found on the *right* half of the A scale, and the correct square root (7.0, 70, or .7) appears directly below on the D scale. The digital part of logarithms for numbers on the C or D scale appears on the L scale, but for numbers less than 1.0 or greater than 10 an integer must be added to the digital part that equals the highest power of 10 contained in the number. Thus the log of 2 is .301, the log of 20 is 1.301 (only the digits .301 appear on the L scale), the log of 2000 is 3.301, the log of 0.2 is $-1 + .301$, usually written as $9.301 - 10$, instead of -0.699, and the log of .02 is $8.301 - 10$. A little practice permits rapid operation, but note that the values can be read to only three digits, at best, on each of the different scales.

Multiplication and division, using the C and D scales, requires moving the slide so that logarithms can be added or subtracted. The C and D scales are identical. The location of any number marked on either was established by making the distance from the left end (or "left index") of the scale to the number proportional to the logarithm of the number. (See Figure 2.1). By moving the slide so that the left (or right) index of the C scale is directly above the value X on the D scale, the product XY can be found on the D scale exactly underneath the value of Y on the C scale. Essentially, the slide rule has added two lengths which correspond to log X and log Y, and the sum, log X + log Y equals log XY. The quotient X/Y can be found by locating X on the D scale, and moving the slide so that the value of Y on the C scale is exactly above X. The quotient X/Y then appears on the D scale beneath the left index of the C scale. (If the slide has to be moved to the left, the quotient appears beneath the right index of the C scale. Note that the quotient Y/X also appears on the slide rule; it is found on the C scale opposite the left or right D index.) The slide rule essentially has subtracted the length of log Y from the length of log X, and the result log X — log Y equals log X/Y. Only the first three digits of a number can be used in most slide rule operations, but the resulting accuracy is adequate for practical purposes.

Figure 2.1. Slide rule operations: squares and square roots; multiplication; division; percentages

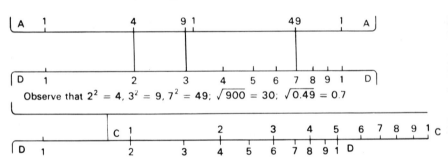

Observe that $2^2 = 4$, $3^2 = 9$, $7^2 = 49$; $\sqrt{900} = 30$; $\sqrt{0.49} = 0.7$

Figure 2.1a. Squares and square roots using A and D scales

Figure 2.1b. Multiplication by 2. The left index of the C scale is set above 2 on the D scale. The value of 2X is read on the D scale beneath the value of X.

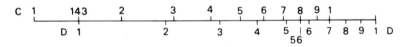

Figure 2.1c. Division. $56 \div 8 = 7$ and $8 \div 56 = .143 = 14.3\%$. When the divisor (denominator) is set on the C scale above the dividend (numerator), the quotient appears under the C index (in this case, the right C index). When the divisor is set on the D scale below the dividend, the quotient appears over the D index (in this case the left index).

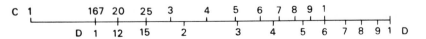

When several percentages must be calculated from the same base, a simple shortcut is possible that reduces the number of slide settings to one or two. If four candidates received 3, 12, 15, and 30 votes, what percentage did each receive? Since the total number of votes cast was 60, move the slide so that the right C index is exactly above the 6 on the D scale, as shown in Figure 2.2a. The required percentages, 5, 20, 25, and 50 percent, can then be read off the C scale directly above the respective values on the D scale. Suppose 50 ballots were divided 5, 8, 12, and 25. Figure 2.2b shows how the respective percentages can

Figure 2.2a.   Cumulative Percentages. Given 3 + 12 + 15 + 30 = 60, find the percentage distribution by setting the C index over 60 (that is, over the line marked 6) on the D scale. The desired percentages then can be read from the C scale above the values on the D scale. Thus 3 = 5%, 12 = 20%, 15 = 25%, and 30 = 50%.

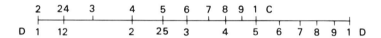

Figure 2.2b.   Cumulative percentages. Given 5 + 8 + 12 + 25 = 50, find the percentages by setting first the right C index above 50 on the D scale. This gives 25 = 50%, 5 = 10%, and 12 = 24%.

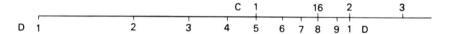

However to find the percentage for 8, the slide must be moved so that the *left* C index is above 50 on the D scale. Then above the number 8 on D appears 16 on C, so 8 = 16%

be read directly off the C scale by first setting the right C index over 5 on D, and then by setting the left C index over 5 on D. Two difficulties must be pointed out. The decimal points in the percentages must be spotted carefully, especially for values less than 1 percent. The third digit for percentages greater than 40 percent is not easy to read, and this may prove frustrating when working with election returns especially. If a large number of percentages must be computed to three digit accuracy, it may prove profitable to invest three or four dollars in a circular slide rule seven inches in diameter with 20 inch C and D scales. These strange looking devices are easy to use, and will always provide three digit accuracy without difficulty. In any case, the accuracy of the percentages can be checked by adding them together to see if they total 100.0 percent. Percentages that add to 99.8, 99.9, 100.1, or 100.2 are also acceptable, for the deviations are probably due to unavoidable rounding errors. If the sum is less than 99.8 percent or greater than 100.2 percent, a mistake has been made.

Three technical points must be kept in mind when working with percentages. Remember that the decimal point is "really" two places to the left of the percentage's decimal point; that is, 75 percent = 0.75, 250 percent = 2.50, and ½ percent = .005. Second, avoid the fallacy of thinking the overall percentage for a group equals the average of the percentages for each category in the group. In 1960 Richard Nixon received an average of 58.0 percent of the vote in Illinois' 102 counties, but of course Chicago voted heavily against him and he received only 49.8 percent of the total vote in the state. The mean percentage for all 102 counties can be found by adding up each of the county values and dividing by 102. The true state percentage can be found only if each county is weighted by the number of votes it cast. That is, multiply the percentage for each county by the total vote for each county, giving the actual number of votes the man received in that county; add up these products, giving his total statewide vote, and divide by the total number of votes cast in the state. This roundabout procedure is, of course, unnecessary if the actual vote breakdown in each county is available. Occasionally, however, the historian's raw data consists of percentages or other averages, and an overall percentage or average has to be computed by the long process. Finally, it is imperative that when percentages are presented in tabular form, the base (the number of cases representing 100 percent) be clearly indicated so that the reader can understand how the percentages were computed, and if he so desires estimate the actual number of cases in each category.

Like any statistical device percentages can be misinterpreted or misused in a way that confuses both reader and researcher. This happens frequently in election analysis when county returns are reduced to percentage form for interpretation, thus obscuring the very relevant point that candidates are elected by pluralities not percentages. Consider the problem of interpreting the relative strength of a third party in three counties, as shown in Table 2.1.

Where was the third party strongest? By percentage analysis, in county M where it garnered 45 percent of the vote. The average resident in M was more likely to encounter a third party man in his neighborhood than citizens in K or L. However, in terms of the internal structure of the party, 4000 (64 percent) of its 6250 supporters live in K, which is likely to house the party's headquarters, convention site, and newspapers. In the legislature, however, the only spokesman for the party was the representative elected in L over a divided opposition. Analysis of the party's strength based on a collective biography of the men it elected to office would, therefore, focus on the peculiar units in which the opposition was closely divided, or where it failed to form a fusion ticket. Note that in terms of the overused concept of "balance of power" the third party was weakest in M, for it could not determine the winner there by shifting all its votes to one of the other parties. If the researcher is particularly interested in the social bases of voting patterns, he will want to explain why the third party won more of the voters in M than in the other two counties. If he is inquiring about the composition of the party's membership, however, he must devote most of his attention to K. The problem can also be acute in two-party contests. Politicians trying to win statewide office will consider an extra percentage point gained in a metropolis far more important than a gain of ten points in a small rural county. Scholars who analyze elections solely in terms

TABLE 2.1   **Hypothetical Election Returns**

COUNTY	THIRD PARTY VOTE	PARTY A VOTE	PARTY B VOTE	TOTAL VOTE
K	4000	4500	1500	10,000
%	40%	45%	15%	100%
L	1800	1600	1600	5000
%	36%	32%	32%	100%
M	450	30	520	1000
%	45%	3%	52%	100%
Totals	6250	6130	3620	16,000
%	39%	38%	23%	100%

of county percentages are giving equal weight to political units, but not equal weight to each vote. They are likely to overlook the bulk of the population and concentrate their research on thinly populated rural areas which may in fact behave differently from metropolitan areas. This constitutes the fallacy of misplaced attention, and is especially dangerous when statistical maps are used. Large, underpopulated rural areas occupy more map space, and thus receive a disproportionate amount of attention than more important compact cities. When dealing with a state like Illinois, where half the population is located in a single county, the best resolution of the fallacy of misplaced attention is to analyze voting patterns in the metropolis separately from the voting patterns of downstate areas. In general the use of percentages removes the population factor from the analysis. In many research designs the population base is irrelevant, for the researcher wants to discover relative rates of occurrence (for example, the percentage of Catholics who are Progressives versus the percentage of Protestants). The question must always be asked and decided whether the use of the percentages confounds rather than aids analysis because of the removal of the population factor.

Distributions of One Variable

Suppose that interval measurements of attribute X are available for all N members of a particular group. The information can be abstracted from the particular individuals who constitute the group, and studied instead as a new attribute that pertains to the entire group itself. This new attribute is the distribution, and three characteristics of the distribution, its shape, average value, and dispersion, suffice for practical purposes to describe the relationship of attribute X to the group. The full body of information about the group can also be summarized visually by a graph or histogram, but it is more convenient for research purposes to deal with shape, average, and dispersion.

Two graphical devices, the histogram and the frequency polygon are commonly used to display a distribution. A histogram is a bar chart whose horizontal axis is marked off in equal intervals of the values of X, and whose vertical axis measures the height of each vertical bar in terms of number of cases or percentages (in the latter case, $100\% = N$). Figure 2.3a shows a simple example. For a group of $N = 10$, the observed values of X are 1.5, 2.0, 2.1, 3.0, 4.0, 4.6, 5.5, 6.6, 7.8, and 9.3. Three intervals are marked along the horizontal axis (or X axis), 0 to 3.99, 4.00 to 7.99, and 8.0 to 11.99, and the bars are drawn in to the proper heights. The frequency polygon consists of the lines connecting the midpoints of the tops of the bars. Note that if different intervals had been marked off along the X axis, the shape of the histogram and frequency distribution would be altered. A few rules of thumb may prove helpful in the construction of these graphs.

1. The intervals should be equal in length (except possibly the last interval on the right).

2. Each interval should be closed on the left and open on the right. Thus in Figure 2.3a the first interval ranges from 0 to 4, including 0 but *excluding* 4.00; the second interval includes 4.00 but excludes 8.00; the third interval includes 8.00 but excludes 12.00.

3. The number of intervals should depend on the number of cases N. When N < 20 (N less than 20), three or four intervals suffice. When 20 < N < 50, use four to six intervals; when 50 < N < 100, use five to eight; when N > 100, use seven or more. Too many intervals produce erratic frequency polygons, while too few result in loss of information.

4.   The vertical axis should extend about 20 percent higher than the highest bar, and it should be about 25 percent shorter than the horizontal axis. Thus if the highest bar is going to be five inches, the vertical axis should extend six inches and the horizontal axis eight inches. If this rule of thumb is followed, a series of graphs will be more nearly comparable, and will be esthetically neater.

Figure 2.3a.   Figure 2.3b.

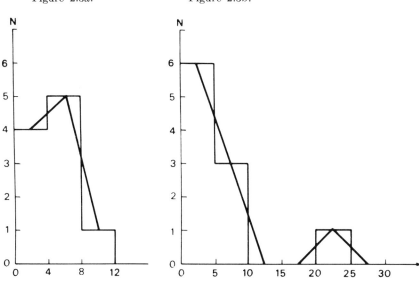

Figure 2.3c.

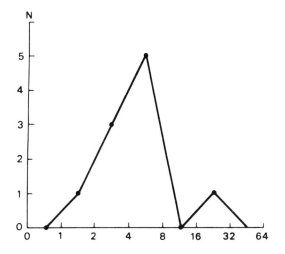

5.   Histograms should be drawn on ordinary graph paper with wide spacing. The bars should touch each other without gaps unless one interval is represented by zero cases. The horizontal axis need not include a zero point, but in every case the vertical axis must include one. "Amputated" histograms result from beginning the vertical scale at a point greater than zero, and distort the shape of the distribution.

6.   When depicting intervals of unequal length, bear in mind that the total *area* of all the bars corresponds to 100 percent of the cases. If two intervals have the same number of cases, therefore, their bars must have equal areas. Hence if one bar is three times as wide as the other, it must be one third as high to have the same area. Figure 2.4a shows an incorrect histogram—the interval from 65 to 85 has only half the number of cases as the interval from 25 to 35, yet is shown with an equal area. Figure 2.4b is the correct form. No matter how different the lengths of intervals, the proper areas can always be found by letting each little square in the bars represent a fixed number of cases. Note also that the width of each bar must be proportional to the numerical length of the interval it represents; that is, the bar representing 65–85 must be twice as wide as the bar representing 55–65. When dealing with transformed horizontal scales, the widths of the bars must be proportional to the transformed intervals (for example, to $\log X$ or X^2 or $\sqrt{X}$).

7.   Outliers, or extremely large values, may require rethinking the purpose of the graph. In Figure 2.3a, for example, if the largest X value were 23.6 instead of 9.3, it would be an outlier. If the outlier represents a quaint but essentially unimportant case (in comparison with the other cases), then the last interval can be open-ended (for example, "8 and over") and include the extreme value. If, however, the fact that one subject has scored a very high value is especially important—for example, if one city has a population of 23.6 thousand and no other city exceeds 8 thousand—then the histogram should be recast so as to emphasize the outlier. This can be done by widening the intervals, say to 0–5, 5–10, 10–15, 15–20, and 20–25, but the result will be badly skewed to the right, as Figure 2.3b shows. To prevent skewness the horizontal axis can be "transformed," with the high end squeezed together and the low end stretched, by using equal length intervals of 1–2, 2–4, 4–8, 8–16, and 16–32. (See Figure 2.3c.) Notice that each interval covers twice as many values as the preceding interval, and each division point is double the preceding point (both criteria are necessary). The use of this transformation permits all the scores to be presented concisely and clearly, without excessive skewness; note also that no change has been made in the basic data.

Once a frequency polygon has been drawn from a histogram, the shape of the distribution can be categorized as flat (or rectangular), symmetric, bell-shaped ("normal"), skewed-right, skewed-left, J-shaped, reverse J-shaped (this type should be transformed), bimodal, or

U-shaped. Idealized examples of each shape appear in Figure 2.5. There are other categories to describe special properties of distribution curves (such as pointedness and roundedness), but the basic shapes will suffice for most purposes of description. Together with a measure of the average and the dispersion, the shape permits the researcher to manipulate a body of numerical data conceptually and helps the reader to visualize the distribution without recourse to a graph. Occasionally when N is very small, the choice of intervals may drastically affect the shape of the frequency polygon. Figure 2.6 shows bimodal, symmetric, and U-shaped distributions for exactly the same data—only the length of the intervals is different. When N is large, the choice of intervals rarely makes much difference.

Important quick inferences can be drawn from the shapes of distributions. A symmetrical shape suggests (it certainly does not *prove*) the existence of a centripetal force pulling all the subjects toward the average, coupled with two or more weaker centrifugal forces that tend to pull the subjects to the left and right. Suppose a report of weekly

Figure 2.4a.   Distribution of U.S. cities over 25,000 population in 1910, by proportion of males over 21 who were Native Born Whites whose parents were also Native Born. (Incorrect Histogram)

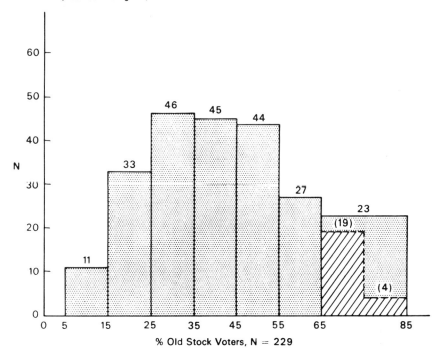

% Old Stock Voters, N = 229

incomes for a category of workers is available, and that the basic hourly wage (the centripetal force) is the same for each man. But because of illness, overtime, and other centrifugal factors that vary the number of hours actually worked, some men will receive more and some less than the average income, thus producing a distribution that may be symmetrical or, if one centrifugal factor is stronger than the other, skewed. If the report on incomes included both high wage skilled workers and low wage unskilled workers, the resulting distribution would be bimodal, with one peak above the normal income for unskilled workers and the other above the corresponding peak for skilled workers. A bimodal shape, therefore, suggests that the group actually consists of two distinct types of subjects who have been lumped together, but who,

Figure 2.4b.   Distribution of U.S. Cities over 25,000 population in 1910, by proportion of males over 21 who were Native Born Whites whose parents were also Native Born. (Correct histogram. Note that 19 cities fell in the 65–75% range, and 4 in the 75–85% range. Fig. 2.4a gives these 23 cities the same area as the 46 cities in the 25–35% range. This is a common mistake when unequal intervals are used. Fig. 2.4b shows the correct areas; the bar for the 65–85% interval, which is twice as wide as the others, is drawn as if it represented 11.5 cases, the average of 19 and 4.) Note that the remainder of the cities' population consisted of women, children, Negroes, immigrants, and sons of immigrants.)

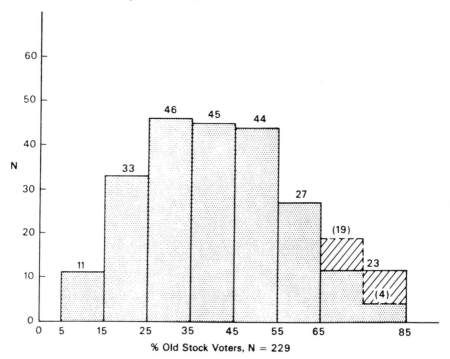

% Old Stock Voters, N = 229

Source Bureau of the Census, *Thirteenth Census of the United States . . . 1910: Abstract* (Washington, 1913)

Figure 2.5.   Standard shapes of distributions

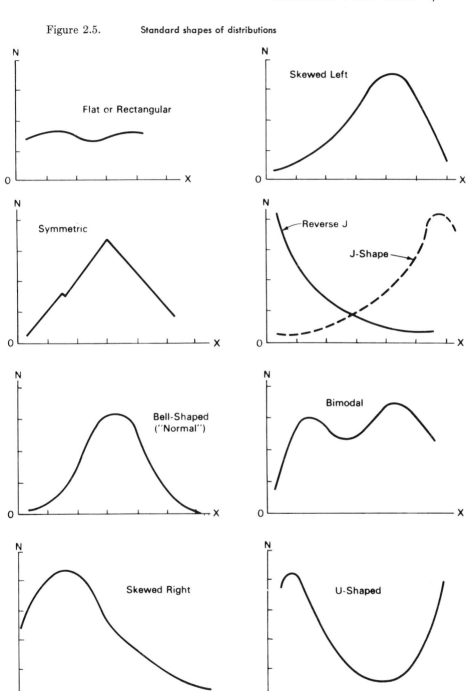

if separated, would display a symmetrical (or skewed) distribution with a single peak.

Distributions of wealth, income, land holdings, and the like for entire communities typically are highly skewed to the right. That is, most people have low or moderate incomes while a very small number have very large incomes. This type of distribution can be considered as an

Figure 2.6. Selection of intervals affects the shape of the distribution: an extreme Example: (Each graph shows the distribution on the following 20 values: 1.6, 2.5, 2.8, 3.2, 3.3, 3.5, 3.7, 3.9, 4.4, 5.6, 6.3, 7.2, 8.1, 8.3, 8.6, 8.8, 9.1, 9.4, 9.7, 11.0)

Figure 2.6a. Using six intervals of length two, the frequency polygon is bimodal.

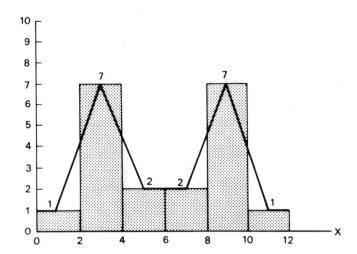

Figure 2.6b. Using four intervals of length three, the frequency polygon is symmetrical.

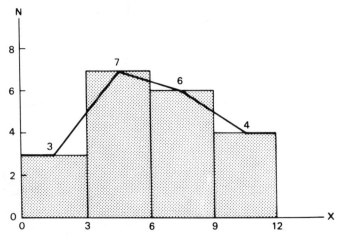

Figure 2.6c.   Using three intervals of length four, the frequency polygon is U-shaped. Note that when more short intervals are used, the total area of the bars is smaller.

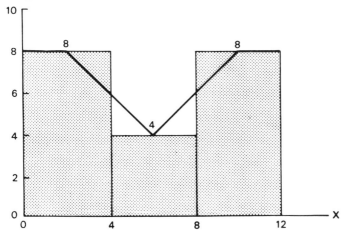

example of inequality, with the rich holding an unequal share of the community's resources. (Chapter 4 contains a detailed discussion of appropriate techniques for measuring inequality.) The distribution of cities by population size is also skewed to the right: many towns have small populations, and a few cities have very large populations. Geographers have spent a good deal of ingenuity constructing models that mathematically produce distributions like these. Figure 2.7a shows right-skewness, when the distribution of Ohio counties is plotted according to the predominance of immigrants in 1910. Figure 2.7b shows the same data in a symmetric graph, thanks to a geometric transformation of the horizontal scale. The intervals form a geometrical progression, 2–4 percent, 4–8 percent, 8–16 percent, 16–32 percent, 32–64 percent, and 64–128 percent. (Of course, the graph is cut off at 100 percent instead of stretching to 128 percent.) This geometrical transformation is the same as a logarithmic transformation, since the values of log 2, log 4, log 8, log 16, log 32, log 64, and log 128 are the equally spaced values .3, .6, .9, 1.2, 1.5, 1.8, and 2.1. In other words, if the *logs* of the X values in Figure 2.7a are plotted, the result is identical with Figure 2.7b.

A U-shaped distribution suggests some sort of all-or-nothing situation, one almost equivalent to the case of a dichotomous variable. For example, the distribution of the variable "Christians as a percent of total population" for 103 countries of the world in the late 1950s was U-shaped. Thirty-five countries scored from 90 to 100 percent Christian, while 38 were under 10 percent. Only 30 countries fell into the very

Figure 2.7a. Ohio counties, 1910, by percentage of males over 21 who were immigrants or sons of immigrants.

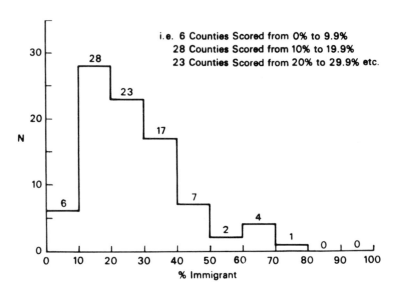

i.e. 6 Counties Scored from 0% to 9.9%
28 Counties Scored from 10% to 19.9%
23 Counties Scored from 20% to 29.9% etc.

Figure 2.7b. Same data, but horizontal scale transformed (geometric or logarithmic scale).

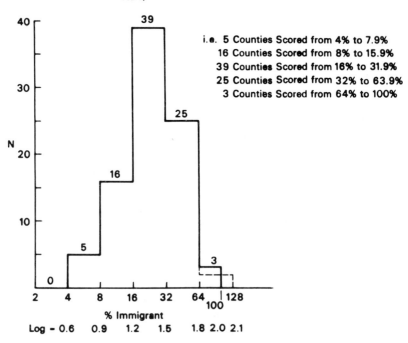

i.e. 5 Counties Scored from 4% to 7.9%
16 Counties Scored from 8% to 15.9%
39 Counties Scored from 16% to 31.9%
25 Counties Scored from 32% to 63.9%
3 Counties Scored from 64% to 100%

wide range of 10–90 percent Christian. Similarly the variables "percent Moslem," and "percent of total vote cast for Communist parties," were clearly U-shaped. For obvious historical reasons, these attributes were of an all-or-none variety, with very few countries taking a middle position.[3]

A J-shaped or reverse-J distribution suggests a strong strain toward unanimity. The distribution of Republican congressmen in terms of their support for a Republican president's legislative program, for example, will show most men with high or very high support scores, but also some with lower scores, thus producing a J-shaped distribution. The distribution of Democratic congressmen on the same variable, however, may produce either a reverse-J (the great majority have low scores), or, in the case of Eisenhower, who bid for bipartisan support, a symmetrical shape.[4]

Averages

An average is a single number that indicates the most typical or representative value in a distribution; it conveys more information about the distribution, in most instances, than the shape or any other statistic, and is therefore the workhorse of all quantitative research. A variety of different averages are in use, but the most important are the *mode*, the *median*, and the arithmetic *mean*. The mode is the most frequent single value (or the midpoint of the interval with the most cases). The mode is not used very often, but in historical studies it can be used to identify the most commonly observed pattern, and is therefore of some interest. If a frequency polygon has one peak, it represents the mode. If there are two (bimodal) or more peaks, the highest one is the mode, but in this case its usefulness in summarizing the distribution is limited. The chief weakness of the mode is that its value often depends on how the intervals are chosen, and that it cannot easily be computed without arraying the data in tabular form. The median is the score of the subject for whom half the subjects score lower and half score higher. (It is *not* the average of the highest and lowest scores, which is called the *midrange*.) The median, unlike the mode, is well defined, and, if N is small, easily found by hand. It is not easy to find by computer, since few canned programs generate its value. The mean, or arithmetic average, written $\bar{X}$ for variable X, is the sum of all the X scores divided by the number of cases N, or $\bar{X} = \Sigma X_i/N$.[5]

In a symmetric distribution the mode, median, and mean are nearly identical. In a distribution skewed to the right (or left), the mean is to the right (or left) of the median, which in turn is to the right (or left) of the mode. These relationships are displayed in Figures **2.8a** and **2.8b**. Interestingly, in a unimodal distribution the distance from the mode to the median is about twice the distance from the median to the mean.

Figure 2.8a. **Right-skew**   Figure 2.8b. **Left-skew**

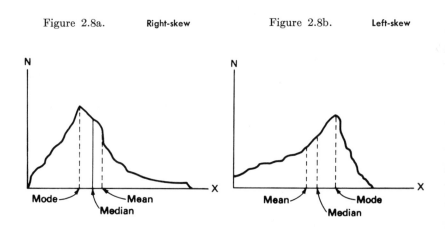

The superiority of the mean over other averages is considerable; only the median is a close competitor. While the technical basis of the superiority of the mean need not concern us, it is important to note that the mean lies closer (in terms of squared distance) to all the other points of a distribution than does any other average. This property becomes very important in the proportional reduction of error approach to measures of correlation, which will be treated in Chapter 3. Note also that if two groups are merged together, the new mean $\bar{X}$ is simply the weighted average of the old means $\bar{X}_1$ and $\bar{X}_2$, or $\bar{X} = (N_1\bar{X}_1 + N_2\bar{X}_2)/(N_1 + N_2)$, where N_1 and N_2 are the sizes of the two subgroups. However, the new median is not a simple function of the two old medians, and indeed it has to be computed anew, although it will lie somewhere between the two old medians. The new mode may be anywhere.

The computation of the mean value when N is large can be troublesome. The straightforward way, using an adding machine or computer, is to add each score together and divide by N. However, two shorter methods are available; they have very good accuracy and save immense amounts of time. Table 2.2 shows the computational procedure for the "shortcut estimate" of the mean, working from tabulated data. The number of cases in each interval N_i is counted—this may be done quickly by hand or derived from a computer tabulation—and the value of the midpoint M_i of each interval noted. Choose the modal interval's midpoint M, in the example, the midpoint 65.0 percent of the interval

60.0–69.9 percent, and subtract M from each M_i. Then multiply N_i times d_i (where $d_i = M_i - M$), and sum up these products, one for each interval, to get $\Sigma N_i d_i$. Note that in this instance $\Sigma N_i d_i$ means summation for each interval, six in all, not summation for each subject, 102 in all. The preceding steps are, in fact, very simple and can be done quickly. The estimated mean is then $M + \Sigma N_i d_i / N$, or in this example, 65.0 percent $+ (-320.4)/102 = 65.0$ percent $- 3.1$ percent $= 61.9$ percent. Actually the true mean for the data was 61.1 percent, so the estimate is quite accurate. The shortcut works because it is reasonable to assume that all 27 values from 50.0–59.9 percent were symmetrically distributed around 55.0 percent, the midpoint of the interval. This assumption, however, is not true for the two extreme intervals. Inspection shows that the average values in these intervals was 34.1 percent and 80.7 percent, so these values are used in the computation instead of 35.0 percent and 85.0 percent. In general, the distribution of points in the extreme intervals is not symmetrical.

TABLE 2.2 Shortcut Estimate of Mean from Tabular Data; X = Republican Percentage of Total Vote Cast for President in 102 Illinois Counties, 1928

INTERVAL # RANGE = i	# COUNTIES IN INTERVAL = N_i	EST. MID-POINT OF INTERVAL = M_i	VALUE OF $M_i - M$ = d_i	VALUE OF $N_i \times d_i$
1 30.0–39.9%	2	34.1%	−30.9%	−61.8%
2 40.0–49.9%	9	45.0%	−20.0%	−180.0%
3 50.0–59.9%	27	55.0%	−10.0%	−270.0%
4 60.0–69.9%	46	65.0% = M	0	0
5 70.0–79.9%	16	75.0%	+10.0%	+160.0%
6 80.0–89.9%	2	80.7%	+15.7%	+31.4%
sum	$N = 102$			$-320.4 = \Sigma N_i d_i$

hence $\bar{X}$ est. mean $= M + \Sigma N_i d_i / N = 65.0\% - 320.4/102 = 65.0\% - 3.1\% = 61.9\%$.

Thus the estimated mean $= 61.9\%$, and long calculation shows that the true mean $= 61.1\%$, indicating only a slight error in estimation.

Source: Richard Scammon, *America at the Polls: 1920–1964* (Pittsburgh: 1965).

A check of the source for Table 2.2 shows, however, that Herbert Hoover won only 56.9 percent of the total Illinois vote, not 61.9 percent or 61.1 percent. Why does the mean differ so much from the statewide figure? Simply because the mean treats its subject (that is, each county) equally, whereas the total gives great weight to Cook county, where Hoover polled only 52.7 percent of the vote.

A second shortcut produces a reasonable estimate of the mean in seconds. Table 2.3 provides a key for its use. The use of this quick estimate will save a great deal of time in preliminary research. For more precise analysis, the first shortcut method should be used; the quick estimate should not be used in published reports.

TABLE 2.3 Quick Estimate of Mean

NUMBER OF SUBJECTS = N	ESTIMATED VALUE OF THE MEAN
3	Use the middle value
4,5,6,7	$\frac{1}{2}$(2nd largest + 2nd smallest values)
8,9,10	$\frac{1}{2}$(3rd largest + 3rd smallest values)
11,12,13,14	$\frac{1}{2}$(4th largest + 4th smallest values)
15,16,17,	$\frac{1}{2}$(5th largest + 5th smallest values)
18,19,20,21	$\frac{1}{2}$(6th largest + 6th smallest values)
22,23,24	$\frac{1}{2}$(7th largest + 7th smallest values)
25,26,27,28	$\frac{1}{2}$(8th largest + 8th smallest values)
29,30,31	$\frac{1}{2}$(9th largest + 9th smallest values)
32,33,34,35	$\frac{1}{2}$(10th largest + 10th smallest values)
36,37,38	$\frac{1}{2}$(11th largest + 11th smallest values)
39,40,41,42	$\frac{1}{2}$(12th largest + 12th smallest values)
43,44,45	$\frac{1}{2}$(13th largest + 13th smallest values)
46,47,48,49	$\frac{1}{2}$(14th largest + 14th smallest values)
50,51,52	$\frac{1}{2}$(15th largest + 15th smallest values)
N	$\frac{1}{2}$($\frac{1}{4}$Nth largest + $\frac{1}{4}$Nth smallest values)

Source: Wilfred J. Dixon and Frank J. Massey, *Introduction to Statistical Analysis* (New York: 1957), p. 265.

Spread, Variance and The Standard Deviation

Measures of central tendency—the averages—are well known and widely used, but the less well known measures of the spread of a distribution play a major role in all advanced statistical work, and are valuable in describing distributions. The simplest measure of how spread out or bunched together is a distribution is the *range,* which is the difference between the largest and smallest value. Since the range involves only two points, it is far less valuable than measures which take all the points into account.

The two distributions shown in Figure 2.9 have the same mean, the same median, the same mode and the same range, yet they are of different shapes; clearly A is more spread out than B which clusters about the mean. That is, the values in B lie closer to the mean, and hence the mean is a better description of B than it is of A. A measure of spread (usually called "dispersion" by statisticians) is necessary to distinguish the two distributions.

Figure 2.9.

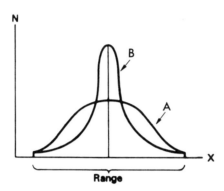

Range

At first glance an obvious measure of spread would be something like $\Sigma(X_i - \bar{X})/N$, the average difference between each point and the mean; this measure, however, equals $\Sigma X_i/N - \Sigma\bar{X}/N = \Sigma X_i/N - N\bar{X}/N = \Sigma X_i/N - \Sigma X_i/N = 0$ by definition of $\bar{X}$, and so is useless. Much better is the average distance from the mean, ignoring positive and negative signs, or $\Sigma|X_i - \bar{X}|/N$. This measure, called the "mean deviation" is good as it goes, but is troublesome to compute and lacks the immense theoretical advantages of the standard deviation.

The basic measure of the spread is the standard deviation (written s) or its square the variance, written s^2. The variance happens to be proportional to the average (mean) *squared* distance between every possible pair of points in the distribution. It is, however, defined precisely as the average (mean) *squared* difference between each point and the mean:

$$\text{variance} = s^2 = \Sigma(X_i - \bar{X})^2/N$$

and

$$\text{standard deviation} = s = \sqrt{\frac{\Sigma(X_i - \bar{X})^2}{N}}$$

To be more precise, if only a *sample* from a larger population is being studied, the s and s^2 values for the sample are those just given, but the best estimates of the standard deviation and variance of the parent population are very slightly larger than these values, and can be found by substituting $N - 1$ in place of N in the above equations. Table 2.4 shows an example of the computation of s and s^2.

TABLE 2.4 Computation of Standard Deviation and Variance: Hypothetical Data

The mean is $\Sigma X_i/N = {}^{49}\!/_7 = 7 = \bar{X}$
The mean deviation is $(3 + 2 + 1 + 0 + 1 + 2 + 3)/N = {}^{12}\!/_7 = 1.7$
The variance is $\Sigma(X_i - \bar{X})^2/N = (9 + 4 + 1 + 0 + 1 + 4 + 9)/N = {}^{28}\!/_7 = 4.0 = s^2$
The standard deviation $= s = \sqrt{4.0} = 2.0$.

$i =$	1	2	3	4	5	6	7
$X_i =$	4	5	6	7	8	9	10
$X_i - \bar{X} =$	-3	-2	-1	0	$+1$	$+2$	$+3$
$(X_i - \bar{X})^2 =$	9	4	1	0	1	4	9

For purposes of calculation the full formula for the variance is inconvenient. Observe that $(X_i - \bar{X})^2 = X_i^2 - 2X_i\bar{X} + \bar{X}^2$, and that $\bar{X}$ is a constant. Therefore, $s^2 = \Sigma(X_i^2 - 2X_i\bar{X} + \bar{X})^2/N = \Sigma X_i^2/N - \Sigma 2X_i\bar{X}/N + \Sigma \bar{X}^2/N$. But $\Sigma 2X_i\bar{X}/N = 2\bar{X}\Sigma X_i/N = 2\bar{X}(\bar{X}) = 2\bar{X}^2$, and $\Sigma\bar{X}^2/N = \bar{X}^2N/N = \bar{X}^2$. Hence $Ns^2 = \Sigma X_i^2 - 2\bar{X}^2 + \bar{X}^2 = \Sigma X_i^2 - \bar{X}^2$, a more convenient form. Other forms are $s^2 = [(\Sigma X_i^2) - N(\bar{X}^2)]/N = \Sigma X_i^2/N - (\Sigma X_i)^2/N^2$. The computation of s^2 and s from ungrouped data requires, therefore, only the computation of ΣX_i and ΣX_i^2. In the data for table 2.4, $\Sigma X_i = 49$, while $\Sigma X_i^2 = 4^2 + 5^2 + \cdots + 10^2 = 16 + 25 + \cdots + 100 = 371$, and $N = 7$. So $s^2 = 371\frac{1}{7} - (49)^2/7^2 = 53 - 49 = 4$. A table of squares (found in all books of statistical tables and in most statistics textbooks) will permit the computation of the variance on an adding machine. The standard deviation, being the square root of the variance, can be found with slide rule or a table of logarithms. Only expensive desk calculators have automatic devices to extract square roots; computers are too time-consuming and expensive to use for simple calculations, unless a very large number of results are needed.

When N is larger than 30, a close estimate of the variance can be made by a shortcut technique very similar to that used to estimate the mean from grouped data. The worksheet should be set up like Table 2.2, but now with a d_i^2 column and a $N_id_i^2$ column. Table 2.5 shows the shortcut method.

TABLE 2.5   Shortcut Computation of Variance and Standard Deviation From Grouped Data

where $M = 25.0 =$ the midpoint of the modal interval.

$$\text{Then variance} = s^2 = \Sigma N_i d_i^2/N - (\Sigma N_i d_i/N)^2 = \frac{17,200}{80} - \left(\frac{420}{80}\right)^2$$

$$= 215.0 - (5.25)^2 = 215.0 - 27.6 = 187.4$$

and $s = \sqrt{s^2} = \sqrt{187.4} = 13.7$.

INTERVAL	MIDPOINT	N_i	d_i	d_i^2	$N_i d_i$	$N_i d_i^2$
0–9.9	5.0	7	−20	+400	−140.0	+2800.0
10–19.9	15.0	12	−10	+100	−120.0	+1200.0
20–29.9	25.0 = M	19	0	0	0	0
30–39.9	35.0	22	+10	+100	+220.0	+2200.0
40–49.9	45.0	14	+20	+400	+280.0	+5600.0
50–59.9	55.0	6	+30	+900	+180.0	+5400.0
Total		N = 80			$\Sigma N_i d_i =$ +420.0	+17,200.0 = $\Sigma N_i d_i^2$

The exact values of the variance and standard deviation involve a fair amount of meticulous calculation. Fortunately it is possible to make fair estimates of the standard deviation in seconds, using one of two shortcuts. As usual, the shortcut estimates are adequate for preliminary work, but should not be used for publication.

For most historical distributions, the range (largest minus smallest value) equals 3 to 5 standard deviations. Hence a rough estimate of s is one fourth the range. If the distribution is highly skewed (J-shaped), or has outliers (one or two values that represent more than 10 or 20 percent of the range), this estimate is too large. Try $\frac{1}{6}$ range or $\frac{1}{7}$ range. If the distribution is U-shaped, the shortcut estimate is too small. Use $\frac{4}{10}$ range or $\frac{1}{3}$ range.

A slightly slower and more accurate estimate of the standard deviation, one that is less dependent on the presence of unusual outliers, involves going in from the end points a little way, and taking one third the range between points 7% from each end. The estimate of s is $\frac{1}{3}$ (V^* - V^{**}), where V^* is the $(.07N + \frac{1}{2})$th largest value, and V^{**} is the $(.07N + \frac{1}{2})$th smallest value. Thus for $N = 50$, $V^* = .07(50) + \frac{1}{2} = 4.0$, or the fourth largest value, and V^{**} is the fourth smallest value. One third of their difference is the standard deviation, roughly. For $N = 150$, $V^* = .07(150) + \frac{1}{2} = 11$, or V^* is the eleventh largest value and V^{**} is the eleventh smallest value.

It is not easy to acquire a feel for the standard deviation the way one obtains a feel for the mean and for different shapes. The next chapter, however, will demonstrate the fundamental importance of the variance in measures of association. Specifically, the approach is to measure the error involved in estimating the shape of distributions. The measure used is the variance—the lower the variance, the closer the values cluster around the mean, and hence the better is the mean as a description of the entire distribution.

The standard deviation plays another important role in theoretical statistics, as part of the theory of the "normal distribution." The normal distribution has a bell-shape, a very complicated algebraic definition, and many elegant, simple mathematical properties. However, it is rarely encountered in historical data and it will not be assumed anywhere in this text unless specifically mentioned. The normal curve is so important that statisticians have measured its exact shape very carefully. Letting the mean equal zero, and a standard deviation equal to one unit on the horizontal scale, results in a normal curve such that 68.3 percent of the curve's area lies within one unit either way of the mean, and 95.5 percent of the area lies within two units either way of the mean. That is, 31.7 percent of the area lies more than one unit from the mean, and 4.5 percent lies more than two units away, while only a tiny 0.27 percent of the area lies more than three units away. Since a normal curve is symmetric, this means that exactly 16.9 percent of the area, which corresponds to 16.9 percent of the values of the distribution, lie more than one standard deviation to the right of the mean, only 2.3 percent lie more than two standard deviations to the right, and only 0.14 percent lie more than 3 s to the right of the mean. (College Board and Graduate Record Exam grades are scored so that the mean is 500 and the standard deviation is 100; using a table of normal curve areas from a statistics textbook, it is possible to estimate what proportion of students, nationwide, will attain scores between 650 and 700, that is, more than 1.5 and less than 2.0 s to the right of the mean.)

FOOTNOTES

1. The definition of levels of measurement depends on the nature of the data. Psychologists, for example, differentiate "interval" and "ratio" levels—and call them all "scales." The distinction is irrelevant in documentary data.

2. For a gloomy review of the inherent inaccuracy of census and economic data see Oskar Morgenstern, *On the Accuracy of Economic Observations* (Princeton: 1963). For a good review of the scope and quality of official statistics in the late nineteenth and early twentieth century in the United States, Canada, India, Australia and 11 European nations,

see John Koren, ed., *The History of Statistics* (New York: 1918). On the failure of the 1960 U.S. Census to reach millions of Negroes, see David M. Heer, ed., *Social Statistics and the City* (Cambridge: 1968).

3. For the detailed statistics, see Bruce Russett, *et al., World Handbook of Political and Social Indicators* (New Haven: 1964), tables 23, 73, 74, 75. The accuracy of the data in this compilation is often suspect (the United States is listed as exactly 55.8 percent Christian, while Sweden scored an amazing 99.0 percent Christian), and some of its computations (especially some standard deviations) are completely wrong. Still, it provides very convenient summaries of a great variety of hard-to-get data, and presents the results of elaborate statistical calculations on the data.

4. DONALD MATTHEWS, *U.S. Senators and Their World* (Chapel Hill, N.C.: 1960), p. 144 for the data. Many fascinating and instructive distributions are plotted in this valuable study.

5. The notation ΣX_i or ΣX used in this book is a short version of the full notation $\sum_{i=1}^{n} X_i$, and, by definition, equals $X_1 + X_2 + X_3 + \cdots + X_n$; unless otherwise specified, the summation extends from X_1 up to and including X_n, where $n = N$, the number of subjects under consideration. The rules for the sigma Σ notation are simple: If A is a constant, $\Sigma A X_i = A X_1 + A X_2 + \cdots + A X_n = A(\Sigma X_i)$, and $\Sigma X_i/A = X_1/A + X_2 A + \cdots X_n/A = (\Sigma X_i)/A$. Thus constants of multiplication and division may be moved from behind the Σ to in front of the Σ. If B is a constant too, then $\Sigma(A X_i + B) = A\Sigma X_i + \Sigma B = A\Sigma X_i + NB$, since $\Sigma B = B + B + \cdots + B$ (that is, B added together N times) $= NB$. However it is *not* true that $\Sigma X_i Y_i = (\Sigma X_i)(\Sigma Y_i)$ when both X and Y are variables, because $\Sigma X_i Y_i = X_1 Y_1 + X_2 Y_2 + \cdots + X_n Y_n$, while $(\Sigma X_i)(\Sigma Y_i) = (X_1 + X_2 + \cdots + X_n)(Y_1 + Y_2 \cdots + Y_n)$. Note also that $\Sigma X_i^2 = X_i^2 + X_2^2 + X_3^2 + \cdots + X_n^2 =$ the sum of the squares of each X value, which is very different from $(\Sigma X_1)^2 =$ the square of the sums of the X values. For example, if $N = 4$, $\Sigma X_i^2 = 1^2 + 2^2 + 3^2 + 4^2 = 1 + 4 + 9 + 16 = 30$, while $\Sigma X_1 = 1 + 2 + 3 + 4 = 10$, and $10^2 = 100$. The sigma notation is used so often in the statistical chapters that the reader should become familiar with it. When in doubt, expand the notation to the full addition of all the values, which makes comprehension a bit easier in especially complicated equations.

MEASUREMENT
OF ASSOCIATION

CHAPTER THREE   The methods presented in this
chapter deal with the discovery,
measurement, and interpretation
of the statistical relationships between two (or more) variables. The
proper technique to be used depends upon the level of measurement of
the two variables.

In general suppose that for a group of N subjects the historian
has obtained for each subject (i) a valid and reliable measure (X_i)
of the subject's score on the variable called X, and a valid and reliable
measure (Y_i) of the subject's score on Y. The problems of validity,
reliability, and distribution of a single variable have already been covered in Chapters 1 and 2. This chapter focuses on the *form, strength,*
and *significance* of the relationship between X and Y.

Form

The form of association for two *interval* variables can be visualized
by a scatter diagram (Figure 3.1) in which the scores on X and Y
for each of the N subjects are represented by the location of points
with respect to the horizontal and vertical axes. Table 3.1 presents the
simple historical data plotted in the scatter diagram. Figure 3.1 illustrates the method used to plot the N points. We use X_i and Y_i to
denote the scores of subject i, and the corresponding point is symbolized
(X_i, Y_i).

Conventionally the horizontal axis is used to locate the X coordinate,
and the vertical axis for the Y coordinate. Usually the "independent"

or "causal" variable, urbanization in this case, is labelled X, and the dependent variable, the one we want to "explain" labelled Y. If there is more than one independent variable, they can be labelled X_1, X_2, X_3, and so on. In that case the value of X_1 for subject 7 is denoted $X_{1,7}$, and in general the value of X_j for subject i is denoted $X_{j,i}$. If more than one dependent variable is involved, different notation and much more advanced methods are called for.

Figure 3.1 reveals a nearly *linear form* for the relationship between X and Y. That is, all the points lie fairly close to the straight line drawn in. The equation for that line (called the "line of best fit") is $Y^* = bX + a$, and the equation is called the "regression equation." Y^* means the "estimated" value of Y, that is the value that would obtain

TABLE 3.1   Democratic Percentage of the Presidential
Vote, 1960, and Percent Urban, 1960: New
England

STATE	X(= % URBAN)	Y(= % FOR KENNEDY)
Connecticut	78.3%	53.7%
Maine	51.3	43.0
Massachusetts	83.6	60.2
New Hampshire	58.3	46.6
Rhode Island	86.4	63.6
Vermont	38.5	41.4

if all the points did in fact lie exactly on a straight line. The coefficients a and b are called "regression coefficients" and contain all the information necessary to draw the line. The coefficient a always equals the value of Y at the point where the line intersects the Y axis, and b is always the slope of the line. (The slope, or tangent of the angle the line makes with the X axis, equals the vertical distance the line rises when X increases one unit.) Figure 3.2 shows the lines that correspond to the regression equations $Y^* = X$, $Y^* = 2X + 3$, $Y^* = -X + 10$, and $Y^* = -.3X + 2$. For the sake of comparison, the line $X = 4$ (which is not a regression equation) is also shown. To draw a line given the a and b coefficients, one needs two points. Obviously one point can be $(0,a)$—that is, the point on the Y axis where $Y = a$. To find another point substitute a convenient value for X in the equation, say $X = 10$, and plot the resulting point $(10, 10b + a)$.

Observe that when b is positive $(b > 0)$, the line always slopes upward from left to right—that is, larger values of X are associated with larger values of Y, and smaller values of X are associated with smaller values of Y. When b is negative $(b < 0)$ the line slopes downward from left to right—that is, larger values of X are associated with smaller values of Y, and vice versa. $b > 0$ indicates positive association; $b < 0$ indicates negative association. When $b = 0$, the line is horizontal and there is no association between X and Y (more precisely, there is no linear association).

The scatter diagram summarizes all the information in a table. The amount of information in the scatter diagram that a regression equation summarizes depends on the strength of association. When the relationship is strong, one can work with the regression equation instead of the entire table.

The coefficients a and b for the best possible straight line, called the "least squares regression coefficients," require a good deal of trouble

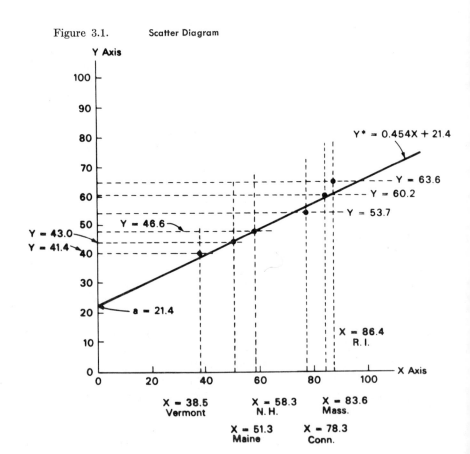

Figure 3.1.   Scatter Diagram

Figure 3.2.   **Regression lines**

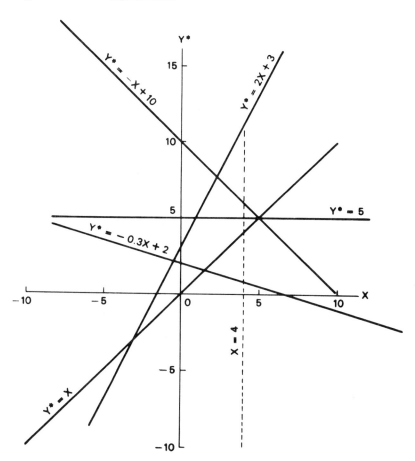

to compute. The formulas are:

$$b = \frac{N \Sigma X_i Y_i - (\Sigma X_i)(\Sigma Y_i)}{N \Sigma X_i^2 - (\Sigma X_i)^2}$$

and

$$a = \frac{\Sigma Y_i - b \Sigma X_i}{N}$$

Fortunately a close approximation to the least squares coefficients can be found by the following quick method:

Step 1: Divide the N subjects into three nearly equal groups according to their X score. Group I, with about $N/3$ subjects, are those with the *lowest* X scores; group III, which must have exactly the same number of subjects as group I, are those with the highest X scores. Group II are the remainder, with X scores all between the highest in I and the lowest in III.

Step 2: Add up the X values for the subjects in group I, and call the sum $\sum\limits_{x}$ I; add up the X values for group III and call the sum $\sum\limits_{x}$ III.

Step 3: Take the subjects in group I and add up their Y scores, and call the sum $\sum\limits_{y}$ I. (Note do *not* divide the subjects into groups on the basis of their Y scores. Groups I, II, and III are based only on X scores.) Then take the sum of the Y scores for group III, and call the sum $\sum\limits_{y}$ III.

Step 4: Then

$$b \approx \frac{\sum\limits_{y} III - \sum\limits_{y} I}{\sum\limits_{x} III - \sum\limits_{x} I}$$

and

$$a \approx \frac{\Sigma Y_i - b\Sigma X_i}{N}$$

One can then calculate Y_i^* by the formula $Y_i^* = bX_i + a$.

Example of quick regression coefficients, using the data in Table 3.1.

GROUP I			GROUP III		
	X	Y		X	Y
Vermont	38.5%	41.4%	Mass.	83.6	60.2
Maine	51.3%	43.0%	R.I.	86.4	63.6
$\sum\limits_{x} I = 89.8$	$\sum\limits_{y} I = 84.4$		$\sum\limits_{x} III = 170.0$	$\sum\limits_{y} III = 123.8$	

$$\Sigma Y_i = 308.5; \quad \Sigma X_i = 396.4;$$

Hence $b \approx \dfrac{123.8 - 84.4}{170.0 - 89.8} = \dfrac{39.4}{80.2} = .49;$

$$a \approx \frac{308.5 - (.49)(396.4)}{6} = \frac{114.3}{6} = 19.05$$

(Note that ΣY_i is the sum of all six Y_i values, and similarly ΣX_i.) The estimated regression equation therefore is

$$Y^* = .49X + 19.1$$

And we can now calculate the estimated Y^* values as follows:

Connecticut	$Y^* = .49(78.3) + 19.1 = 57.4\%$	true Y	$= 53.7\%$
Maine	$.49(51.3) + 19.1 = 44.2$	"	$= 43.0$
Massachusetts	$.49(83.6) + 19.1 = 60.0$	"	$= 60.2$
New Hampshire	$.49(58.3) + 19.1 = 47.6$	"	$= 46.6$
Rhode Island	$.49(86.4) + 19.1 = 61.4$	"	$= 63.6$
Vermont	$.49(38.5) + 19.1 = 37.9$	"	$= 41.4$

The estimated and true Y values do not perfectly coincide, but the fit is especially good for historical data.

Strength of Association (Correlation)

The measure of the strength of association, or "correlation" between interval variables X and Y is simply a measure of how close the estimated Y^* values from the regression equation are to the actual Y. If the error is small, X and Y are strongly correlated; if the error is large, X and Y are weakly correlated, or uncorrelated. If X and Y are strongly correlated and b is positive, the correlation is positive; if b is negative, the correlation is negative. If X and Y are weakly correlated, b makes little difference.

We need to measure the total error of estimation of Y^* but quick inspection shows that $\Sigma(Y_i - Y_i^*)$ will not do, for this sum can be zero even if the line is a very bad fit. (In fact the sum is zero if only the line passes through (M_x, M_y), the point corresponding to the two means.) However it is a *necessary condition* for the *best* line that it make the sum zero, in which case $0 = \Sigma(Y - Y^*) = \Sigma Y - \Sigma Y^* = \Sigma Y - \Sigma(bX + a) = \Sigma Y - b\Sigma X - Na$ hence $a = (\Sigma Y - b\Sigma X)/N$. The most useful definition of total error of estimate is the sum of the squared errors, $\Sigma(Y - Y^*)^2$.

The strength of association between X and Y (where X and Y may be interval, categorical, ranked, dichotomous, or a mixture of any two) can be simply and elegantly defined as the *proportionate reduction of error* in estimating Y^* knowing X versus not knowing X. Let error E_1 be the total error of estimate $\Sigma(Y - Y^*)^2$ *not* using X to get Y^*, and let error E_2 be the total error $\Sigma(Y - Y^*)^2$ using X to get Y^*. Then the strength of association between X and Y is simply

$$R^2 = \frac{E_1 - E_2}{E_1} = 1 - \frac{E_2}{E_1}$$

R^2 is called the "coefficient of determination" in the statistics literature, and seldom receives the attention from users of statistics it deserves.

A variety of correlation coefficients appear in the statistics literature; instead of reviewing all of them, we will concentrate on those especially valuable for historical research. Specifically we will consider the "Pearson product moment correlation coefficient" usually designated r, and used when both X and Y are interval variables; the "Spearman-rho" coefficient, written r_s and used when both X and Y are ranked variables; the "mean square contingency" coefficient phi, written ϕ and used when both X and Y are dichotomous variables; Pearson's "coefficient of contingency" C and the Goodman-Kruskal lambda-b, written λ-b, when X and Y are both categorical; and the "correlation ratio" eta used when X is categorical and Y is interval, ranked, or dichotomous. We will also give quick methods for estimating r based on "tetrachoric" correlations r_{tet}.

The common correlation coefficients, r, r_s, and ϕ all have the following properties: the square of the coefficient ranges from 0 to 1, and the coefficient itself ranges from -1 to $+1$; when the coefficient is near $+1$ or near -1, the variables are very strongly correlated, and it is possible to predict the Y values very well using the X values. When the coefficient is near 0, the variables *may* possibly be strongly associated, but they are not associated in a linear form. (The association may be curved, or in the case of dichotomous variables, a third "control" variable Z may bring out a hidden association between X and Y).

By more precisely stating the proportional reduction of error (PRE) definition, the formulas for all the correlation coefficients can be easily derived, and we will show the derivation of r. Note that *no* assumption regarding the distribution of X and Y is necessary in the definitions. Specifically there is no need to specify that X and Y are normally distributed. (There are certain elegancies when X and Y are in fact normally distributed, and the sampling reliability of the coefficients is precisely known in that case; but the historian need not assume normality!) The PRE criteria are:

a) State a rule for estimating each Y_i^* not knowing the values on X. (In the case of r the rule is simply $Y^* = $ mean Y or $Y_i^* = \dfrac{\Sigma Y}{N}$. In the case of λ-b the mode is used instead of the mean.)

b) State a rule for estimating Y_i^* given knowledge of X_i. (In the case of r the rule is simply the least squares regression equation $Y_i^* = bX_i + a$; in the case of λ-b the modal value of Y in each category of X is used. In multiple regression and correlation the rule is $Y_i^* = a + b_1X_{1i} + b_2X_{2i} + \cdots + b_kX_{ki}$.)

c) Define the total error of estimate for rules a) and b). For r, E and ϕ the rule used is error $= \Sigma(Y_i - Y_i^*)^2$.

d) Then $R^2 = \dfrac{\text{total error from a)} - \text{total error from b)}}{\text{total error from a)}} = \dfrac{E_1 - E_2}{E_1}$

The correlation coefficient is the square root of R^2.

We can now derive the formula for r from the PRE criteria and some relatively easy algebra.[1] For rule a) we estimate Y^* simply by the mean value of Y, $Y_i^* = \bar{Y}$. For rule b) we use the regression formula that gives the best fit, that is, the equation $Y_i^* = bX_i + a$ where b and a are such as to maximize R^2. Since $0 = \Sigma(Y - Y^*)$, as we have noted, $a = \bar{Y} - b\bar{X}$. For notational simplicity, let us write $y_i = Y_i - \bar{Y}$ and $x_i = X_i - \bar{X}$. For step c) we note that the total error from step a) is $\Sigma(Y_i - \bar{Y})^2 = \Sigma y_i^2$. The total error from step b) is $\Sigma(Y_i - Y_i^*)^2 = \Sigma(Y_i - bX_i - a)^2 = \Sigma(Y_i - \bar{Y} + b\bar{X} - bX_i)^2$ (this is true by the definition of a) $= \Sigma[Y_i - \bar{Y} - b(X_i - \bar{X})]^2 = \Sigma(y_i - bx_i)^2 = \Sigma y_i^2 - 2b\Sigma y_i x_i + b^2\Sigma x_i^2$ (this is true by expanding the squared term). To minimize the error of step b) we must minimize the last expression by choosing a suitable value of b. Simple calculus tells us to differentiate the last expression with respect to b, and set the result equal to zero. This procedure gives $-2\Sigma y_i x_i + 2b\Sigma x_i^2 = 0$, hence $b = \Sigma y_i x_i / \Sigma x_i^2$ which is the definition of b. Now the total error of step b) is

$$\Sigma(y_i - bx_i)^2 = \Sigma y_i^2 - 2b\Sigma x_i y_i + b^2\Sigma x_i^2$$

$$= \Sigma y_i^2 - \frac{2(\Sigma x_i y_i)}{\Sigma x_i^2}\Sigma x_i y_i + \frac{(\Sigma x_i y_i)^2}{(\Sigma x_i^2)^2}\Sigma x_i^2$$

$$= \Sigma y_i^2 - \frac{2(\Sigma x_i y_i)^2}{\Sigma x_i^2} + \frac{(\Sigma x_i y_i)^2}{\Sigma x_i^2} = \Sigma y_i^2 - \frac{(\Sigma x_i y_i)^2}{\Sigma x_i^2}$$

Hence

$$R^2 = \frac{\Sigma y_i^2 - \Sigma y_i^2 + \dfrac{(\Sigma x_i y_i)^2}{\Sigma x_i^2}}{\Sigma y_i^2} = \frac{(\Sigma x_i y_i)^2}{\Sigma x_i^2 \Sigma y_i^2} \qquad \text{qed.}$$

and

$$r = \sqrt{R^2} = \frac{\Sigma x_i y_i}{\sqrt{\Sigma x_i^2}\sqrt{\Sigma y_i^2}}$$

Since $y_i = Y_i - \bar{Y}$ and $x_i = X_i - \bar{X}$ we have the familiar formula:

$$r = \frac{\Sigma(X_i - \bar{X})(Y_i - \bar{Y})}{\sqrt{\Sigma(X_i - \bar{X})^2}\sqrt{\Sigma(Y_i - \bar{Y})^2}}$$

In practice the computation of r is very tedious by hand, even for a small N. It goes faster on a desk calculator, especially those large models that can produce all the sums, cross-products and squares in one set of operations. In general the easiest formula to use for computing purposes is

$$r = \frac{N\Sigma X_i Y_i - (\Sigma X_i)(\Sigma Y_i)}{\sqrt{[N\Sigma X_i^2 - (\Sigma X_i)^2][N\Sigma Y_i^2 - (\Sigma Y_i)^2]}}$$

and

$$b = \frac{N\Sigma X_i Y_i - (\Sigma X_i)(\Sigma Y_i)}{N\Sigma X_i^2 - (\Sigma X_i)^2}$$

If for shorthand we call $\Sigma X_i = A$, $\Sigma Y_i = B$, $\Sigma X_i Y_i = C$, $\Sigma X_i^2 = D$ and $\Sigma Y_i^2 = E$, we have

$$b = \frac{NC - AB}{ND - A^2} \qquad a = \frac{B - bA}{N}$$

and

$$r = \frac{NC - AB}{\sqrt{(ND - A^2)(NE - B^2)}}$$

One of the problems in computing r is that $ND \approx A^2$ and $NE \approx B^2$ (although one can *never* have $ND < A^2$ nor $NE < B^2$), hence the computations should not be rounded—the multiplications should use at least four digits in the final product. The following example from the data in Table 3.1 shows the computational procedures.

Here $A^2 = 157132.96$, $B^2 = 95172.25$, $AB = 122289.40$, $NC = 127494.30$, $ND = 168585.84$, $NE = 97723.26$, $NC - AB = 5204.90$, $ND - A^2 = 11452.88$, $NE - B^2 = 2551.01$, $b = .4543$, $a = 21.40$, $R^2 = .930$, and $r = .965$.

STATE	X_i	Y_i	X_i^2	Y_i^2	$X_i Y_i$
Me.	51.3	43.0	2631.69	1849.00	2205.90
Vt.	38.5	41.4	1482.25	1713.96	1593.90
N.H.	58.3	46.6	3398.89	2171.56	2716.78
Mass.	83.6	60.2	6988.96	3624.04	5032.72
R.I.	86.4	63.6	7464.96	4044.96	5495.04
Conn.	78.3	53.7	6130.89	2883.69	4204.71
	$A = 396.4$	$B = 308.5$	$D = 28097.64$	$E = 16287.21$	$C = 21249.05$

Using the values of the regression coefficients a and b just obtained, the values of Y_i^* can be calculated without great trouble. Since the correlation coefficient $r = .965$ is very high, we expect that the estimated Y_i^* values should be very close to the true Y_i.

STATE	X_i	$bX_i = .454X_i$	$Y^* = .454X_i$ $+21.4$	TRUE Y_i	ERROR $Y_i - Y_i^*$
Me.	51.3	23.3	44.7	43.0	-1.7
Vt.	38.5	17.5	39.8	41.4	$+2.5$
N.H.	58.3	26.5	47.9	46.6	-1.3
Mass.	83.6	38.0	59.4	60.2	$+0.8$
R.I.	86.4	39.2	60.6	63.6	$+3.0$
Conn.	78.3	35.6	57.0	53.7	-3.3
			$\Sigma 308.5$	$\Sigma 308.5$	$\Sigma 0.0$

Some cautions should be noted. In the interpretation of r, we can only say that the Kennedy vote followed very closely, in a *linear form*, the percent urbanized in the New England states. The result may or may not be true for other states, or for the *counties* that constituted New England. The focus here is on the relationships between variables for states, not for counties or for individual voters. To conclude that city people must have voted more heavily for Kennedy than did rural people in New England, while hinted at by the high correlation, does not in fact follow. Such a conclusion is an example of the "ecological fallacy" which derives conclusions about individual behavior from state (or county) data.

Every computer center has programs to compute r and the regression coefficients a and b. The user should be careful if he has missing data (for example, if the percent urban for Maine is missing, the computer will treat it erroneously as zero unless a "missing data program" is used). For more detailed analysis, the researcher will be interested in the error term, usually called the "residual from regression," for it shows how much and in what direction each state differed from the linear pattern.

To avoid the tedium of calculation one should start with "tetrachoric correlations" instead of the Pearson product moment coefficient of correlation. For preliminary pattern searching the tetrachoric coefficients written r_{tet} are all that need be calculated, and with a little practice one correlation between two variables over 40 or 50 subjects can be calculated in less than five minutes. When all the correlations among several variables X, Y, Z, . . . must be found, r_{tet} can be calculated in less time than it takes to drive to the computer center.

Two methods of calculating r_{tet} are available. The short method should be used when the number of subjects N is even, and when they can be divided at the median into an upper and lower range. If N is odd, or the division at the median is impossible or misleading, the long method can be used, and it still is very fast. To work the short method, find the median value for the subjects on variable X and on variable Y. Mark each subject that is *above* the median on X; and mark each subject that is above the median on Y. Using a worksheet like Table 3.4, a check or circle can be placed in the X column for each subject that is above the median on variable X. Then simply count the number of subjects that are above the median on *both* X and Y, and call this sum A. A equals the number of subjects that were checked in both the X and the Y column. To find r_{tet} first divide A by N. If A/N is greater than .25, the correlation is *positive* and the value of r_{tet} can be found opposite the corresponding value of A/N in Table 3.2. If A/N is less than .25, the correlation is *negative*, and the value of r_{tet} can be found in the table opposite the value of $.50 - (A/N)$. Although r_{tet} is usually close to the Pearson r, and also ranges from

TABLE 3.2 Values of r_{tet} Corresponding to Obtained Values of A/N when X and Y are both divided at the median

A/N or $.5 - A/N$	r_{tet}	A/N or $.5 - A/N$	r_{tet}
.250	.000	.390	.771
.260	.063	.400	.809
.270	.125	.410	.844
.280	.187	.420	.876
.290	.249	.430	.905
.300	.309	.440	.930
.310	.368	.450	.952
.320	.426	.460	.969
.330	.482	.470	.982
.340	.536	.480	.992
.350	.588	.490	.998
.360	.637	.500	1.000
.370	.685		
.380	.729		

−1 to +1, r_{tet} usually does not exactly equal r. Nor can one get the regression coefficients a and b from the computation of r_{tet}.

Frequently it is difficult or impossible to divide one or more variable exactly at the median. For example, suppose one variable is percent Republican vote, but the exact voting returns are lost and the historian only knows whether the Republican or the Democratic candidate won. If there were, say, 30 wards and the Republican candidates were elected in 20 and Democrats in 10, then the 20 high and 10 low wards on percent Republican are known. Tetrachoric correlation may be used— and, in fact, no other way of estimating r is possible since the precise returns are lost! In this case divide the subjects into high and low on X, and high and low on Y anyway. Count the number that are high on both (call this A), the number that are low on both (call this D), the number high on X but low on Y (call this C) and the number low on X but high on Y (call this B). Then compute A times D and B times C. If AD is greater than BC, the correlation is positive and r_{tet} can be found by looking for the value of BC/AD in Table 3.3. If BC is greater than AD the correlation is negative, and r_{tet} can be found by looking for the value of AD/BC in Table 3.3 and adding a minus sign.

Example of the computation of r_{tet} for the data in Table 3.4: The median percent Whig in 1840 was 53.8, and the median percent Whig in 1844 was 48.7. Since $N = 25$ states, an odd number, use the longer method of computation, and the values in Table 3.4.

A equals the number of states that were *above* the median in *both* 1840 and 1844. They were Connecticut, Delaware, Georgia, Kentucky, Massachusetts, North Carolina, Ohio, Rhode Island, Tennessee, and Vermont. Thus $A = 10$.

B equals the number of states that were *above* the median in 1840, but *below* the median in 1844. They were Indiana and Louisiana, so $B = 2$.

C equals the number of states that were *below* the median in 1840, but *above* in 1844. They were Maryland and New Jersey, so $C − 2$.

D equals the remainder of the states, so $D = 11$. Hence $AD = 10 \times 11 = 110$, $BC = 2 \times 2 = 4$, $AD/BC = (110/4) = 27.5$, and $BC/AD = 4/110 = .0364$. Use the smaller value, .0364 with a plus sign on the coefficient. Entering .0364 in Table 3.3 shows that r_{tet} lies between +.880 and +.866. Interpolation gives $r_{tet} = +.876$. The value of the Pearson r for the same data is $r = +.820$, but a good hour's worth of work on a desk calculator was required to find and check it!

Tetrachoric correlations are useful when the amount of random error in the original data is very large. Random errors of observation and measurement depress the value of r below its true value, but have only a slight effect on r_{tet}. Note that r_{tet} should *not* be used when X and Y are really dichotomous variables. In that case use ϕ, and save r_{tet} for those problems in which both X and Y are interval variables, at least theoretically.

A more general measure of association than r, and one that is much easier to compute, is Spearman's rho, written r_s. Like other correlation coefficients r_s ranges from $+1$ (perfect agreement) to zero (no association) to -1 (perfect inverse agreement.) Spearman's rho is *not*, however, a measure of how well a straight line fits the points in a scatter diagram. Instead it is a measure of "monotonic" form. A relationship is perfectly monotonic if Y always increases when X increases, or when Y always decreases when X increases. All the lines in Figure 3.3a are monotonic, but none of those in Figure 3.3b are monotonic. Clearly if a relationship is linear it is monotonic, but the reverse does not hold. To compute r_s only the relative ranks of the N subjects on each variable are necessary. Sometimes interval data are available, and can quickly be transformed from ranks, with each subject having a rank from 1 (low) to N (high). In some cases the rankings are all the data available. If two subjects are tied, they should be given an average rank. Thus if two subjects are tied for fifth and sixth place, each should be assigned the rank 5½. If three subjects are tied for seventh, eighth, and ninth place, each should be assigned the rank eighth. For each subject it is necessary to subtract the Y rank from the X rank, and square the difference. Thus if one

Figure 3.3a.   Monotonic curves   Figure 3.3b.   Non-Monotonic curves

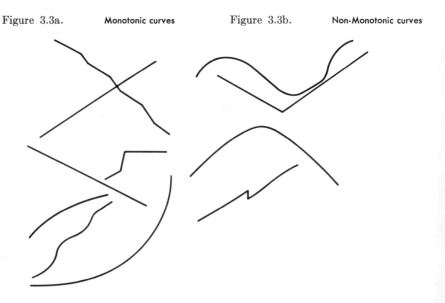

TABLE 3.3 Values of r_{tet} corresponding to obtained values of BC/AD or AD/BC (whichever is smaller), when X or Y or both are not divided at the median

For both tables, the value of r_{tet} for other values of $A/N, .5 - A/N, BC/AD$ or AD/BC may be found by simple interpolation. All values were computed on the basis of the formulas

$$r_{tet} = \sin \left[\frac{360° A}{N} - 90° \right]$$

and

$$r_{tet} = - \cos[180°/(1 + \sqrt{BC/AD})].$$

BC/AD or AD/BC	r_{tet}	BC/AD or AD/BC	r_{tet}
.000	1.000	.225	.531
.005	.979	.250	.500
.010	.959	.275	.471
.015	.942	.300	.443
.020	.925	.325	.417
.025	.909	.350	.392
.030	.894	.375	.369
.035	.880	.400	.346
.040	.866	.450	.305
.045	.853	.500	.266
.050	.840	.550	.231
.060	.815	.600	.198
.075	.780	.700	.139
.085	.759	.750	.113
.100	.728	.800	.090
.125	.682	.850	.068
.150	.639	.900	.041
.175	.601	.950	.020
.200	.565	1.000	.000

subject scores tenth on variable X, and seventh on variable Y, the difference is three, and the difference squared is $+9$. These squared differences, one for each of the N subjects, are added together to get ΣD^2. Then

$$r_s = 1 - \frac{6\Sigma D^2}{N(N^2 - 1)}$$

There are no regression coefficients that go with Spearman correlations.

TABLE 3.4   Calculation of r_s: X = percent Whig of total vote by states in 1840 and Y = percent Whig vote by states in 1844

Here $r_s = +.86$ indicating a very strong relationship between the Whig vote in 1840 and the Whig vote in 1844, as was to be expected. Note that South Carolina and states admitted after 1840 are excluded. The Whig percentages by state may be found in Svend Peterson, *A Statistical History of the American Presidential Elections* (New York: 1963), pp. 25–30.

STATE	X % WHIG 1840	X RANK IN 1840	Y % WHIG IN 1844	Y RANK IN 1844	$D = Y - X$	D^2
Alabama	45.6%	4	40.9%	4	0	0
Ark.	43.3	1	36.6	2	1	1
Conn.	55.4	16	50.8	18	2	4
Del.	55.0	15	51.2	19	4	16
Georgia	55.8	18½	48.8	14	4½	20¼
Illinois	48.8	5	42.4	5	0	0
Indiana	55.8	18½	48.4	11	7½	56¼
Kent.	64.2	25	54.1	23	2	4
Louis.	59.7	22	48.7	13	9	81
Maine	50.1	8	40.5	3	5	25
Maryland	53.8	13	52.4	21½	8½	72¼
Mass.	57.4	20	51.4	20	0	0
Mich.	51.7	10	43.7	8	2	4
Miss.	53.5	12	43.3	7	5	25
Missouri	43.6	2	43.0	6	4	16
N.H.	44.6	3	36.3	1	2	4
N.J.	51.8	11	50.5	17	6	36
N.Y.	51.2	9	47.9	10	1	1
N. Caro.	57.9	21	52.4	21½	½	¼
Ohio	54.1	14	49.7	15	1	1
Penn.	50.0	7	48.6	12	5	25
R.I.	61.2	23	59.6	25	2	4
Tenn.	55.6	17	50.1	16	1	1
Vermont	63.9	24	54.9	24	0	0
Virginia	49.2	6	46.9	9	3	9

$$\Sigma D^2 = 369$$

$$r_s = 1 - \frac{6\Sigma D^2}{N(N^2 - 1)} = 1 - \frac{6 \times 369}{25 \times 624} = 1 - \frac{2214}{15,600} = 1 - .14 = +.86.$$

The computation of r_s is illustrated in Table 3.4. Appendix B is a convenient table of squares from which one can find the square of the differences without trouble. To save the small amount of calculation necessary in the formula for r_s use the "nomograph" in Appendix C, with which the value of r_s can be found very quickly once ΣD^2 is known.

Spearman's rho is often used carelessly to estimate r to save time. Since a linear relationship is monotonic, two variables with a high r will have a high r_s and a high r_{tet}. In the example for the Whig votes in 1840 and 1844, $r_{tet} = +.88$, $r_s = +.86$, and the Pearson $r = +.82$. However if the relationship between the two variables is monotonic but not linear, then r_s will be greater than r. In fact it is possible to have a high r_s but a r close to zero for the same data. Thus r_s should be used as an approximation to r only when a linear pattern appears on the scatter diagram. However, r_s is a perfectly good measure of the strength of (monotonic) association in its own right, and its interpretation and value do not depend on its closeness to r. For many historical problems of pattern-seeking there is no reason that r should be preferred to r_s, and of course r_s has an immense advantage in ease of computation. Note, however, that r_{tet} is even easier to compute than r_s. Like r_s, r_{tet} is an approximation of r when r is high. When r is close to zero, however, r_{tet} appears to be more closely related to the measure of monotonicity r_s. If the number of cases is small, or if r_s is close to zero, the historian should conclude that there is no monotonic relationship between the two variables. There may be other curved relationships, however. As a convenient rule, the value of r_s^2 should exceed $1/(N-1)$ if one wants to conclude that a monotonic relationship holds.

The appropriate measure of strength of association for two dichotomous variables is the "mean square contingency coefficient" phi, written ϕ. To compute ϕ first set up a two-by-two (or "fourfold") table as shown in Table 3.5. The four cell entries (a, b, c, d) indicate the count of the number of subjects that score 'yes' on both X and Y $(= a)$, 'no' on X and 'yes' on Y $(= b)$, 'yes' on X and 'no' on Y $(= c)$, and 'no' on both $(= d)$. The marginal values, $a + b$, $c + d$, $a + c$, and $b + d$, are simply the sum of the respective rows and columns. The total number of subjects $N = a + b + c + d$.

TABLE 3.5 Two-by-Two Table

			3.5a Raw data variable X				3.5b Percentage $(N = 100\%)$ X		
			YES	NO			YES	NO	
Variable	Y	yes	a	b	$a+b$	yes	p_{11}	p_{12}	p_{1-}
		no	c	d	$c+d$	no	p_{21}	p_{22}	p_{2-}
		Σ	$a+c$	$b+d$	N	Σ	p_{-1}	p_{-2}	1.00

The Pearson product-moment r can be calculated for a two-by-two table by scoring 'yes' responses as $+1$ and 'no' responses as zero. It turns out with some simple algebra that $r = \phi$. The full formula is:

$$\phi = \frac{ad - bc}{\sqrt{(a + c)(b + d)(a + b)(c + d)}}$$

Note that ϕ ranges from $+1$ to -1, and has the same sign as the difference of cross products $ad - bc$. Only when $b = c = 0$ does $\phi = +1$, and only when $a = d = 0$ does $\phi = -1$. When $ad = bc$, $\phi = 0$, and X and Y are independent. If the cell entries in the two-by-two table are each divided by N, the percentage two-by-two table that results is shown by Table 3.5b. Then

$$\phi = \frac{p_{11}p_{22} - p_{12}p_{21}}{\sqrt{p_{-1}p_{-2}p_{1-}p_{2-}}}$$

Computation is facilitated by Appendix A which gives the values of $\sqrt{p_{1-}p_{2-}}$ and $\sqrt{p_{-1}p_{-2}}$ for all percentages from 0 to 100.

If the four marginal totals are roughly the same (say, none is more than twice as large as any other), then a short-cut approximation to ϕ is

$$\phi \approx \frac{a}{a + c} - \frac{b}{b + d} \quad \text{or} \quad \phi \approx \frac{a}{a + b} - \frac{c}{c + d}$$

The last two new expressions are the difference of proportions across the table $a/(a + c) - b/(b + d)$ and down the table $a/(a + b) - c/(c + d)$ and they correspond to the b regression coefficients in Pearson correlations.

Statistics texts often mention the use of chi-square and Yule's Q in connection with two-by-two tables. Chi-square, written χ^2, is simply $N\phi^2$, and is *not* a measure of strength of association. Rather, chi-square is a test of whether or not the "true" ϕ in a total population equals zero, given the observed two-by-two table representing a sample drawn from the total population. Yule's Q is easily computed from the formula $Q = (ad - bc)/(ad + bc)$. Although Q ranges from $+1$ to -1 like a correlation coefficient, its interpretation is rather circuitous. For most purposes ϕ should be used instead of Q. (Probably the popularity of Q comes from its easy computation, and the fact that in absolute value it is always larger than ϕ.)

Often all the information wanted from a two-by-two table is provided by the difference in proportions $a/(a + c) - b/(b + d)$. If Y is the dependent variable, the difference of proportions suggests the importance X plays in changing the observed frequency of 'yes' responses to Y. The greater the difference, the more important is the role of variable X. If the difference of proportions is zero, then the same proportion of subjects who score 'yes' on X score 'yes' on Y as the proportion of those who score 'no' on X and 'yes' on Y. In other words, X has no visible effect on the proportion of 'yes' responses on variable Y. When the difference of proportions is zero, $\phi = 0$ no matter what the marginal totals are. Table 3.6 illustrates the calculation of ϕ for a roll call of considerable historical interest.

TABLE 3.6 Vote of Commons to Repeal the Corn Laws, May 15, 1846

Either a slide rule or logarithms or a desk calculator can be used in the computation. By slide rule, the proportions are $114/355 = .322$ and $235/245 = .960$, so the approximation method gives $\phi \approx .322 - .960 = -.637$, a very accurate approximation indeed. See William O. Aydelotte, "The Country Gentlemen and the Repeal of the Corn Laws," *English Historical Review*, 82 (1967), 54.

| | | X = "PARTY" AFFILIATION | | |
		CONSERVATIVES	"LIBERALS"	Σ
	aye	114	235	349
Vote	nay	241	10	251
	Σ	355	245	600

$$\phi = \frac{114 \times 10 - 241 \times 235}{\sqrt{349 \times 251 \times 355 \times 245}} = \frac{-55,500}{87,300} = -.636$$

The "correlation ratio," eta, written E^2, is a versatile coefficient of correlation that should be used when X, the independent variable, is categorical, and the dependent variable Y is dichotomous, ranked, or interval. Although we will not derive the formulas for E^2 from the PRE criteria, such a derivation is based on rule a) $Y_i^* = $ overall Y mean $= \bar{Y}$; rule b) $Y_i^* = $ the mean Y value of the X category in which subject i falls $= \bar{Y}_i$; and rule c), error $= \Sigma(Y - Y^*)^2$.

Case 1. Y is dichotomous.

$$E^2 = 1 - \frac{\Sigma N_i p_i q_i}{N p q}$$

where N_i is the number of subjects in the ith X category, p_i = number of 'yes' scores/N_i = percent of subjects in X category i who score 'yes' on Y, and q_i = number of 'no' scores in category i/N_i. Note that $p_i + q_i = 1.00$ for each X category. As an example in Table 3.7 we use Aydelotte's analysis of the 'Conservative' voting patterns on the motion to repeal the Corn Laws, taken in the House of Commons on May 15, 1846; here $E = +.37$.

TABLE 3.7

To compute E^2 add the $N_i p_i q_i$, $\Sigma N_i p_i q_i = 16.3 + 12.1 + \cdots + 2.3$
$$= 66.5$$

Hence

$$E^2 = 1 - \frac{N_i p_i q_i}{Npq} = 1 - \frac{66.5}{77.2} = 1 - .861 = .139$$

and

$$E = \sqrt{E^2} = \sqrt{.139} = .37.$$

	VOTE: Y VARIABLE					
CONSTITUENCY CATEGORY (X-VARIABLE)	AYE	NAY	N_i	p_i	q_i	$N_i p_i q_i$
England & Wales, counties and universities	19	117	136	.14	.86	16.3
England & Wales, boroughs with under 500 voters	19	33	52	.37	.63	12.1
England & Wales, boroughs with 500–999 voters	30	21	51	.59	.41	12.3
England & Wales, boroughs with over 1000 voters	25	34	59	.42	.58	14.4
Scotland & Ireland, counties and universities	12	33	45	.27	.73	9.1
Scotland & Ireland, boroughs	9	3	12	.75	.25	2.3
All constituencies	114	241	355	.32	.68	77.2

The value of E^2 can be also computed directly, without first finding the proportions of yes and no responses in each category. Let (yes$_i$) = the number of yes responses in X category i. And let T be the total number of yes responses. Then

$$E^2 = \frac{\sum \dfrac{(\text{yes}_i)^2}{N_i} - \dfrac{T^2}{N}}{T - \dfrac{T^2}{N}}$$

Complicated as the formula looks, its calculation is simple. For the data from Table 3.7,

$$\sum \frac{(yes_i)^2}{N_i} = \frac{19^2}{136} + \frac{19^2}{52} + \frac{30^2}{51} + \frac{25^2}{59} + \frac{12^2}{45} + \frac{9^2}{12}$$

$$= \frac{361}{136} + \frac{361}{52} + \frac{900}{51} + \frac{625}{59} + \frac{144}{45} + \frac{81}{12}$$

$$= 2.7 + 7.0 + 17.6 + 10.6 + 3.2 + 6.7$$

$$= 47.8$$

$$T = 114, \frac{T^2}{N} = \frac{114^2}{355} = \frac{12,996}{355} = 36.6$$

Hence

$$E^2 = \frac{47.8 - 36.6}{114.0 - 36.6} = \frac{11.2}{77.6} = .144$$

and

$$E = \sqrt{E^2} = \sqrt{.144} = .38$$

Note that the two methods of calculating E^2 give slightly different results (.139 and .144), which is due entirely to rounding errors.

The interpretation of E^2 differs somewhat from the interpretation of r and r_s. Since the X categories are not in any particular order, there is no linear or monotonic pattern to uncover. The data in the example can be thought of in terms of the overall pattern in which 114 out of 355 men, or 32 percent, chose to vote for repeal of the Corn Laws. In the six different categories the proportion of men who voted for repeal ranged from 14 to 75 percent. Clearly the categories, in this case type of constituency, did make a difference, for if the category made no difference, the proportion of aye votes in each would be close to the 32 percent overall average. On the other hand, the categories did not completely determine voting patterns, for if they had either 100 percent or 0 percent of the men in each category would have voted aye. The strength of the association between categories and votes, therefore lies somewhere between 0 and 1. Using $E = \sqrt{E^2} = .38$ locates the strength of the particular categories used. Of course, other ways are possible to categorize the Members of Parliament, and Aydelotte gives a number in his article. One can dichotomize the Members into Businessmen and Nonbusinessmen. When the independent variable is also dichotomized, $E^2 = \phi^2$. Of the 41 Businessmen, 13 (or 32 percent) voted aye; of the 314 Nonbusinessmen, 101 (or 32 percent) voted aye. The difference in proportions $[= a/(a+c) - b/(b+d)]$ in terms of the corresponding two-by-two table, is zero, hence $\phi = E = 0$. One can dichotomize the

Members according to whether or not they attended a public school. Of the 198 men who did not, 56 (or 36 percent) voted aye. The difference in proportions is $.29 - .36 = -.07 \approx \phi$. Hence $\phi^2 \approx (-.07)^2 = .005 = E^2$. Thus this dichotomy is not particularly strong, and indeed is much weaker than the one used in the example. One can compare the strength of categories by taking their ratio of E^2 values. Thus the E^2 coefficient can and should be used when one wishes to summarize statistically the strength of association between two variables when the independent variable is categorical.

Case 2. X is categorical and Y is ranked. A very convenient formula is available for the calculation of E^2 when Y is a ranked variable ranging from 1 to N. (Either the 1 or the N may represent the highest Y value.) Simply find the sum T_i of the Y ranks of the N_i subjects in category i. Compute T_i^2/N_i for each category, and add to obtain

$$\sum \frac{T_i^2}{N_i}$$

Then

$$E^2 = \frac{12 \sum \dfrac{T_i^2}{N_i} - 3N(N+1)^2}{N^3 - N}$$

Note that $(N - 1)E^2 = H$, where H is known as the "Kruskal Wallis H statistic" and can be used to find the probability that $E^2 = 0$ for a population when the observed E^2 is calculated from a sample of subjects drawn randomly from that population.

Case 3. X is categorical and Y is interval. The derivation of the formula for E^2 can be shown when Y is interval. Using the PRE criteria, rule a) put $Y^* =$ the overall mean of $Y = \bar{Y}$; rule b), put $Y_i^* =$ the mean Y value for the X category in which i falls $= \bar{Y}_i$. Then rule c) gives: error $1 = \Sigma(Y - T/N)^2$ (where $T = \Sigma Y$ for all N subjects) and error $2 = \Sigma(Y - T_i/N_i)^2$ (where T_i is the sum of the Y values for the N_i subjects in the ith X category). Then by rule d):

$$E^2 = \frac{\sum \left(Y - \dfrac{T}{N}\right)^2 - \sum \left(Y - \dfrac{T_i}{N_i}\right)^2}{\sum \left(Y - \dfrac{T}{N}\right)^2}$$

$$= \frac{\Sigma Y^2 - 2\dfrac{T}{N}\Sigma Y + \sum \dfrac{T^2}{N^2} - \Sigma Y^2 + 2\Sigma Y \dfrac{T_i}{N_i} - \sum \dfrac{T_i^2}{N_i^2}}{\Sigma Y^2 - 2\dfrac{T}{N}\Sigma Y + \sum \dfrac{T^2}{N^2}}$$

Now

$$\Sigma Y = T$$

$$\sum \frac{T^2}{N^2} = N \frac{T^2}{N^2} = \frac{T^2}{N}$$

and

$$\Sigma Y \frac{T_i}{N_i} = \sum^{k} \frac{T_i^2}{N_i}$$

where $\sum^{k}$ means summation only over the k different X categories, and

$$\sum^{k} \frac{T_i^2}{N_i^2} = \sum^{k} \frac{T_i^2}{N_i}$$

By substituting these equivalent values into the PRE definition of E^2 one obtains the following simplified formula, which can be used for computation:

$$E^2 = \frac{\sum^{k}\left(\frac{T_i^2}{N_i}\right) - \frac{T^2}{N}}{S - \frac{T^2}{N}}$$

where $S = \Sigma Y^2$.

To put the computation procedure simply, for each of the k X categories, separately, add up the Y scores of the N_i subjects in the category to get T_i. Then add up all the k values of T_i to get the overall sum of the N subjects' Y scores, T. Then, square *each* of the N subjects' Y scores separately, and add these squares to get $\Sigma Y^2 = S$. Then square each of the T_i values, and divide by the corresponding N_i. Add up these k quotients to obtain $\sum^{k} T_i^2/N_i$. Finally compute the value of T^2/N. Then plug all these values into the computing formula for E^2 above, and the answer pops out. It looks like a great deal of trouble, but it somehow is rather easy in practice. Example, hypothetical data. Suppose we have the electoral success scores for four categories of counties. We want to find out the strength of the association between county type and success score. The observed values and calculation appear in Table 3.8.

If the mean Y values for each X category are known, ($= \bar{Y}_i$) as well as the number of subjects in each category (N_i), the formula for E^2 can be written:

$$E^2 = \frac{\sum^{k} N_i(\bar{Y}_i)^2 - \frac{T^2}{N}}{S - \frac{T^2}{N}}$$

It may happen that a historical source, such as a compendium of census results, gives only the mean and number for each category, and does not reproduce the individual scores. The historian can not deduce $S = \Sigma Y^2$ from such scanty data. He may be able, however, to find the primary source material (say, the manuscript census reports) and sample a small number of subjects. He does not have to use the X categories in sampling; all that is sought is an estimate of S. Now the denominator divided by N,

TABLE 3.8

Since

$$T = 950, \frac{T^2}{N} = \frac{950^2}{17} = \frac{902,500}{17} = 53,088$$

And

$$S = 60^2 + 55^2 + 65^2 + 50^2 + 45^2 + \cdots + 40^2 + 35^2 = 58,100$$

Hence

$$E^2 = \frac{53,612 - 53,088}{58,100 - 53,088} = \frac{524}{5,012} = .104$$

and

$$E = \sqrt{.104} = .32.$$

X CATEGORY	Y SCORES	N_i	T_i	$\frac{T_i}{N_i} = \bar{Y}_i$	$\frac{T_i^2}{N_i}$
Wheat-growing	60, 55, 65	3	180	60.0	10,800
Cotton-growing	50, 45, 30, 90	4	215	54.0	11,556
Corn & Hogs	80, 75, 40, 60	4	255	64.0	16,256
Ranching	55, 60, 25, 85, 40, 35	6	300	50.0	15,000
		17	950	56.0	53,612

or $[S - (T^2/N)]/N$ equals the *variance* of the distribution of Y. Using quick techniques to estimate the standard deviation, one has: $(N - 1)$ (est st. dev.)$^2 \approx S - T^2/N$. This estimate of the denominator may be used without serious difficulty. If no practical method of estimating the standard deviation is available, all is not lost. Probably the reason the historian wanted E^2 is to compare it with the E^2 produced by another X variable, that is another categorization of the population. The X variable with the largest value of $\Sigma N_i(\bar{Y}_i)^2$ or its equivalent $\Sigma T_i^2/N_i$ has the largest

E^2. And the ratio of the E^2 coefficients for two different X variables is:

$$\frac{E^2}{E'^2} = \frac{\Sigma N_i(\bar{Y}_i)^2 - \dfrac{T^2}{N}}{\Sigma N'_i(\bar{Y}_{i'})^2 - \dfrac{T^2}{N}}$$

Now T^2/N is the same for both X variables, but the number of categories, the N_i and the $\bar{Y}_i$ differ.

A typical research problem is the comparison of the strength of relationship among pairs of variables when all the variables are categorical. A variety of measures exist for this situation, none of which are fully satisfactory. With the exception of lambda-b, to be discussed below, the measures are also troublesome to compute. Fortunately most computer centers have library programs designed to print-out the cross-tabulation of two categorical variables and compute several measures of association.

Suppose X has three categories and Y has four. Tables 3.9a and 3.9b show two possible cross-tabulations (or "contingency tables") with $N = 100$.

TABLE 3.9a   Hypothetical Contingency Table

CATEGORIES		X VARIABLE			
		a	b	c	Σ
	a	5	3	2	10
Y	b	10	6	4	20
	c	15	9	6	30
	d	20	12	8	40
	Σ	50	30	20	100

TABLE 3.9b   Hypothetical Contingency Table

CATEGORIES		X VARIABLE			
		a	b	c	Σ
	a	10	0	0	10
Y	b	0	0	20	20
	c	0	30	0	30
	d	40	0	0	40
	Σ	50	30	20	100

Despite the fact that the marginal totals are the same for both tables, the relationships between X and Y are very different. In fact, the relationship is zero in Table 3.9a, and very nearly perfect in Table 3.9b (actually it is as perfect as possible, given the fact that the number of categories differs). To prove that X and Y are independent in Table 3.9a, try rewriting the table in percentage terms, so that each column adds up to 100 percent. Then rewrite the table so that each row adds up to 100 percent. The columns (or rows) will then be identical for Table 3.9a, and very dissimilar for Table 3.9b.

The problem is to devise a measure of strength of association similar to ϕ for the 2×2 table that will yield the value zero for Table 3.9a, and be close to one for Table 3.9b. The recommended measure is Pearson's "coefficient of contingency," C, which equals the square root of chi-square divided by chi-square plus N. Unfortunately, C does *not* meet the *PRE* criteria. Chi-square (written either with the Greek symbol χ^2 or the plain X^2) is a widely used statistic computed from contingency tables, and used primarily to decide if the sample size is large enough to justify the conclusion that a real relationship between the variables holds for the target population. The historian will have occasion to use chi-square only when he is worried that his sample might be too small, and its value is usually computed by library programs. To calculate C, however, chi-square is needed. For hand calculation the following procedure will produce X^2 and C without excessive trouble, although it is advisable to use a desk calculator so the intermediate numbers will have at least four significant digits.

1. Take each cell entry in the 4×3 (or, generally, $r \times c$) table separately.
2. Square the cell value, giving v^2.
3. Divide v^2 by the column sum (for the column in which the cell is located), giving $v^2/$col Σ.
4. Divide $v^2/$col Σ by the row sum (for the row in which the cell is located), giving $v^2/($col $\Sigma \cdot$ row $\Sigma)$.
5. Add together the values obtained in step 4 for each of the $r \times c$ cells, giving a sum $S = \Sigma \ v^2/($col $\Sigma \cdot$ row $\Sigma)$.
6. Then chi-square $= X^2 = N(S-1)$, where N, of course, is the total number of subjects in the table.
7. Finally, $C = \sqrt{X^2/(X^2+N)} = \sqrt{(S-1)/S}$ qed.

The technique can be illustrated for Table 3.9b. Steps 1 and 2 produce the numbers shown in Table 3.10a; step 3 produces the numbers shown in Table 3.10b; step 4 produces the numbers shown in Table 3.10c. Step 5, the summation of the cell values in Table 3.10c, gives $S =$

TABLE 3.10

TABLE 3.10a			TABLE 3.10b			TABLE 3.10c		
Values of v^2			Values of $\dfrac{v^2}{\text{col } \Sigma}$			Values of $\dfrac{v^2}{\text{col } \Sigma \cdot \text{row } \Sigma}$		
100	0	0	2	0	0	.2	0	0
0	0	400	0	0	20	0	0	1.0
0	900	0	0	30	0	0	1.0	0
1600	0	0	32	0	0	.8	0	0

$.2 + .8 + 1.0 + 1.0 = 3.0$. Hence chi-square equals $100(3.00 - 1) = 200$, and C equals $\sqrt{(S-1)/S} = \sqrt{2/3} = \sqrt{.67} = .82$ as desired. Note that C is less than 1.0; this results from the peculiar nature of C itself. The maximum value of C in a square 3×3 table is .82, in a square 4×4 table .87, and so on up to a square 10×10 table, .95. Other measures have been designed as substitutes for C which do have maximum values of 1.0, most notable "Tschuprow's T," which is included in many computer programs. Unfortunately, T is next to useless, for in most cases its values are very small, and not at all comparable to ϕ in a 2×2 table. The historian who spots a C value above .7 can be very happy, for the relationship can clearly be seen in the full contingency table. If the sample size N is small, there is a chance that an observed value of C is misleading, and that $C = 0$ in the target population. Table 3.11 uses chi-square to find how likely this unhappy result is

TABLE 3.11   Chi-Square Test of Significance for Contingency Tables

(upper right = 90% confidence level; lower left = 95% level)

NUMBER OF COLUMNS (OR ROWS)

		2	3	4	5	6	7	8	9	10
	2	2.7 / 3.8	4.6	6.3	7.8	9	11	12	13	15
	3	6.0	7.8 / 9	11	13	16	19	21	24	26
Number of	4	7.8	13	15 / 17	19	22	26	30	33	37
rows (or	5	9	16	21	24 / 26	28	33	38	43	47
columns)	6	11	18	25	31	34 / 38	40	46	52	58
	7	13	21	29	36	44	47 / 51	54	61	68
	8	14	24	33	41	50	58	62 / 66	70	78
	9	16	26	36	46	56	65	74	79 / 84	88
	10	17	29	40	51	62	72	83	93	98 / 103

If the true contingency coefficient for a population is $C = 0$, a small random sample may nevertheless show a value of $C > 0$. This table shows the value of chi-square (X^2) corresponding to the 10 percent and 5 percent probabilities that the true $C = 0$. Find the two entries (one in the upper half, one in the lower half) corresponding to the number of rows and columns in the table, and either compute X^2 or take its value from a computer printout. If X^2 is greater than the upper entry, the chances are less than 1 in 10 that true $C = 0$. If X^2 is greater than the bottom entry, the chances are less than 1 in 20 that true $C = 0$. Thus for a 5×7 table, the critical value of X^2 at the 90 percent confidence level is 33, and at the 95 percent level 36. The value of N (sample size) is not needed for using this table, but it does enter the computation of X^2.

When *both* X and Y are ordered categorical variables, several other measures of association are available, of which the most popular is Goodman and Kruskal's "gamma." Its calculation is tedious unless computer programs are available, or if only a handful of gammas are required.[2] Unlike C, gamma ranges from -1 to $+1$. Unfortunately, the two measures are not comparable. That is, if gamma for two ordered variables V and W is $+.7$, and C for two unordered categorical variables X and Y is $+.5$, it does not follow that the strength of association in the first case is higher than in the second. When gamma and C are both computed for the same table (both variables are ordered), the value of gamma is higher than the value of C. And for 2×2 tables, the value of gamma = Yule's $Q = (ad - bc)/(ad + bc)$ which is usually much higher than ϕ. While C lacks a PRE definition, gamma has one, but it is so complicated that it can not be neatly interpreted. On the other hand, lambda–b has a very simple (perhaps too simple) PRE definition.

When the historian has a large number of contingency tables he can sort through them by looking for high values of C. If the computer program available does not calculate C, it is possible to substitute chi-square *on the condition that the value of N is approximately the same for all the tables.* In that case, the ranking of the tables in order of chi-square will be the same as the ranking in order of C. The various shortcomings of the contingency coefficient and its substitutes should not be too upsetting. These measures are best used to discover the contingency tables that contain the strong relationships and which, therefore, may deserve special attention. Further analysis of the tables involves conversion of the entries into percentages, with $N = 100$ percent or each row sum = 100 percent or each column sum = 100 percent. Patterns can then be spotted quickly.

The last traditional measure of the strength of association we will describe is Goodman and Kruskal's lambda–b, written λ–b. It can be used when both X and Y are categorical variables, and is the easiest of all coefficients to compute. Table 3.12 illustrates a typical problem it can handle well, that of summarizing the relationship between "section" (the X variable) and "position on slavery" (the Y variable.) Each of the 227 representatives in the first session of the 26th Congress (1840) was classified by the section of the country he came from, and also by his scale position on slavery roll-call votes. The slavery categories are ranked from 1 = proslavery to 6 = anti-slavery, but λ–b makes no use of the fact that the categories are ranked.

TABLE 3.12   Section and Slavery categories for the 227
Representatives in the 26th Congress
(1840)*

		NEW ENG.	MID- ATL.	MID- WEST	SOUTH ATLANTIC	SOUTH CENTRAL	Σ
		X = SECTION					
	1	1	4	1	*35*	*15*	56
	2	0	11	7	10	6	34
Y	3	7	20	*9*	8	13	*57*
score	4	2	5	2	2	3	14
	5	6	*22*	5	0	1	34
	6	*20*	8	4	0	0	32
	Σ	36	70	28	55	38	227

* Adapted from Thomas B. Alexander, *Sectional Stress and Party Strength* (Nashville: 1967), p. 33.

Suppose now that we try to estimate Y^* for a congressman chosen at random, and that we know neither his voting record nor his section. The best estimate is $Y^* = 3$, for the Y category 3 has 57 men in it, more than any other Y category, and is therefore the modal Y category. We can expect to have a probability of 57/227 of assigning the random congressman to the correct Y category, and the probability of $(227 - 57)/227$ of error. Now suppose we are given the information about the X category for each man. If we know our random congressman is from New England, the best estimate of Y^* is now $Y = 6$, the modal Y category for the New England column. Similarly if we know the man is from the Mid-Atlantic section, we estimate $Y^* = 5$, for 5 is the modal Y category for his column. Similarly we assign a Midwesterner to $Y^* = 3$, and a South Atlantic or South Central man to $Y^* = 1$. The probability of a correct assignment, given X, is therefore $(20 + 22 + 9 + 35 + 15)/227 = 101/227$ and the probability of error, knowing X, is $(227 - 101)/227$. Now we make use of the PRE criterion for R^2 and we have:

$$R^2 = \lambda\text{-}b = \frac{(227 - 57) - (227 - 101)}{(227 - 57)}$$

$$= \frac{101 - 57}{227 - 57} = \frac{44}{227} = .258$$

To make λ-b comparable to the other coefficients (which equaled $\sqrt{R^2}$) we have $\sqrt{\lambda\text{-}b} = \sqrt{.258} = .51$.

Lambda–b equals zero when the modal Y category is the same for each X category. Lambda–b is $+1$ when all the men in each column are in one Y category, as shown by the following examples. The value of $\sqrt{\lambda\text{-}b}$ ranges from 0 to $+1$ instead of from -1 to $+1$ because there is no meaning to a negative relationship. Notice that the columns, and the rows, can be arranged in any convenient order without affecting the value of λ-b.

EXAMPLES of λ-b

TABLE 3.13   Republican Representatives in the 34th Congress (1856) by section and slavery scale position*

In this case $\lambda\text{-}b = \dfrac{(22 + 27 + 30 + 0 + 0) - 79}{102 - 79} = \dfrac{79 - 79}{23} = 0.$

		X CATEGORY					
		NE	MA	MW	SA	SC	Σ
	1–4	0	0	0	0	0	0
	5	0	0	1	0	0	1
Y	6	3	16	3	0	0	22
scale	7	22	27	30	0	0	79
	Σ	25	43	34	0	0	102

* Alexander, *Sectional Stress and Party Strength*, p. 93.

Although there were in fact slightly different distributions of Y scores for each X category, the modal Y category was always 7, so the value of λ-b is zero.

TABLE 3.14 Hypothetical Data

In this example $\lambda-b = \dfrac{(15 + 12 + 8 + 14 + 12) - 20}{61 - 20}$

$$= \dfrac{61 - 20}{61 - 20} = 1.$$

		NE	MA	MW	SA	SC	Σ
				X CATEGORY			
	1	0	0	0	14	0	14
	2	0	0	0	0	12	12
Y	3	0	12	8	0	0	20
category	4	0	0	0	0	0	0
	5	15	0	0	0	0	15
	Σ	15	12	8	14	12	61

The coefficients C, r, r_s, and ϕ are symmetrical, that is, they have the same value regardless of whether X or Y is the independent variable. The regression coefficients a and b, however, are not symmetrical; they assume that X is the independent variable. By assuming that Y is the independent variable, and X is dependent, we can compute new regression coefficients a' and b'. It is interesting to note that $bb' = r^2$.

The coefficients E^2 and $\lambda-b$ are not symmetrical. In the case of $\lambda-b$, the correlation coefficient $\lambda-a$ can be computed when X is the dependent variable simply by using modal rows instead of modal columns in the calculation. A symmetrical lambda coefficient λ that lies between $\lambda-a$ and $\lambda-b$ can be formed by adding together the numerators in the formulas for $\lambda-a$ and $\lambda-b$, and dividing that sum by the sum of the two denominators. The statistics literature contains no versions of a coefficient corresponding to E^2 when the dependent variable is categorical. However the PRE criteria permit us to invent one.

Suppose then that X, the independent variable is interval, and Y the dependent variable is categorical. For example, we have 100 New Yorkers from the 1780s of whom 25 were Tories in the Revolution, 25 were Neutrals, 25 were Federalist patriots and 25 were Antifederalist patriots. Their political category is the dependent variable Y, and we wish to measure the strength of association with Y of the interval variable age. In other words, we want to see if older men invariably were Tories, and younger men Federalists, or exactly what the form and strength of the relationship was. Not knowing a man's age (X score) we assign him at random to one of the four groups so that the probability of his being put in any one category equals the relative size of that category. (Note that we do not put him in the modal Y category, and the lambda coefficient did.) Then if we know a man's age, we assign him to the category whose mean age is the closest to his age. If a man was 34 years old in 1780, and the Federalists averaged 35 years in age while the Antifederalists averaged 30 years of age, we assign the man to the Federalist category. This approach is intuitively straightforward, and the correlation coefficient that results is not hard to compute. Before we can define the correlation which we will call theta, θ, however, we need to define the total error (rule (c) of the PRE criteria). We define the total error as the number of men placed in the wrong category. Then

$$R^2 = \theta = \frac{\text{error } 1 - \text{error } 2}{\text{error } 1}$$

To compute θ find the mean age of all the men in each category. Then arrange the categories in order from youngest (on the average) to oldest. Say the youngest group was the Antifederalists with mean age 30. The rule will classify as Antifederalist anyone who is closer to 30 years old than he is to 35, the mean age of the next group. Thus anyone under 32-$\frac{1}{2}$ years of age will be classified Antifederalist. Count the number of men in this age interval who were in fact Antifederalists, and call this number A_1. If the neutrals' mean age was 40, the rule will call Federalist anyone closer to age 35 than to age 30 or 40, or in other words anyone in the range from 32-$\frac{1}{2}$ to 37-$\frac{1}{2}$ years. Count the actual number of Federalists in this age group, and call the total A_2. If the Tories' mean age was 50, the rule will classify as neutral anyone from 37-$\frac{1}{2}$ years to 45 years of age, and anyone over 45 years as a Tory. Count the actual number of neutrals and Tories respectively in these two age intervals, and call the totals A_3 and A_4. The number of errors by knowing X therefore is simply $N - (A_1 + A_2 + A_3 + A_4)$.

The number of errors in classifying the men without knowing their ages works out to error $1 = N - (1/N)\Sigma N_i^2$ where N_i as usual is the number of men in category i. By a little algebraic manipulation we have:

$$\theta = \frac{N\Sigma A_i - \Sigma N_i^2}{N^2 - \Sigma N_i^2}$$

If the categories do not overlap in ages, that is if everyone in the youngest category is younger than everyone in the older categories, and if everyone in the second youngest category is younger than everyone in the older categories, and similarly for the other categories, then the correlation is $+1$. That is, by knowing a person's age we could categorize him exactly. If the distribution of ages happens to be especially perverse, the rule for assignment knowing age may be inferior to random assignment, in which case θ will have a value less than zero. Such distributions are highly unlikely in history.

Example, hypothetical data. Suppose we have $N = 10$, of whom three were Tories (aged 40, 50, and 60), three were Antifederalists (aged 25, 30, and 35), and four were Federalists (aged 20, 25, 30, and 35). The Federalists were the youngest category, with mean age $27\text{-}\frac{1}{2}$ years, then came the Antifederalists with mean age 30, and then the Tories, with mean age 50. Classify everyone in the interval from 20 to $(27\text{-}\frac{1}{2} + 30)/2 = 28\text{-}\frac{3}{4}$ as Federalist, and note that there were $A_1 = 2$ Federalists in this group. Classify everyone from age $28\text{-}\frac{3}{4}$ to $(30 + 50)/2 = 40$ as an Antifederalist, and note that $A_2 = 2$ men have been accurately categorized. Everyone from 40 to 60 should be classified Tory, and since one Tory was exactly 40, this group has $2\text{-}\frac{1}{2}$ men accurately classified. Thus $\Sigma A_i = 6\text{-}\frac{1}{2}$. The values of N_i are 3, 3, and 4; hence $\Sigma N_i^2 = 9 + 9 + 16 = 34$. And since $N = 10$, $N^2 = 100$. Hence

$$\theta = \frac{10 \times 6\text{-}\frac{1}{2} - 34}{100 - 34} = \frac{65 - 34}{66} = .47$$

Multiple Regression Analysis

Multiple regression analysis, the core of econometrics, extends the simple two variable regression procedure to include one dependent and K independent variables.[3] The object is to find the values of the constants b_1, b_2, ... b_k (called "partial regression coefficients") and a, which provide the greatest reduction of error in predicting the values of the dependent variable Y using the linear form:

$$Y^* = a + b_1X_1 + b_2X_2 + \cdots + b_kX_k$$

The multiple correlation coefficient, R^2, is simply the proportionate reduction of error, $1 - \Sigma(Y - Y^*)^2/\Sigma(Y - \bar{Y})^2$.

The estimated or predicted value of Y for each of the N subjects can be found by substituting the observed values of the X variables into the linear form.

Except in the case of two independent variables, the computation of the partial regression coefficients is so tedious that library computer programs should be used. The input in these programs consists of the observed values of Y, X_1, X_2, ... X_k for each of the N subjects. The output consists of the partial regression coefficients, the value of R^2, and a variety of other useful statistics, including the means and standard deviations for each variable, the matrix of correlation coefficients for each pair of variables, and "F" and "t" tests of sampling significance. These tests usually are unnecessary in historical research.[4]

The values of b_1, b_2, . . . b_k can be interpreted in much the same way as the b coefficient in two-variable regression. If X_1 changes by 10%, say, *and all other X_i variables remain unchanged*, then the value of Y^* increases by $(.10)b_1$. For example, suppose $N = 75$ counties, Y = percent Republican of total vote, X_1 = per capita income (in dollars), X_2 = percent urban, and X_3 = region (a dichotomous variable with value 1 for the eastern part of the state, and 0 for the rest), and suppose multiple regression analysis produces values of $a = 30.5$, $b_1 = .003$, $b_2 = -.12$, and $b_3 = 25$. Then the linear form becomes:

$$Y^* = 30.5 + .003X_1 - .12X_2 + 25X_3$$

This means that a county with $X_1 = \$2000$, $X_2 = 40\%$, and $X_3 = 1$ has an estimated Y value of $30.5 + (.003)(2000) + (-.12)(40) + 25 = 30.5 + 6.0 - 4.8 + 25.0 = 56.7$ percent Republican. If for this county the observed $Y = 58.0$, the error is $Y - Y^* = 58.0 - 56.7 = 1.3$. Notice that the Republican percentage increases 3 points for every increase of $1000 in per capita income, declines by 1.2 percent for every increase of 10 percent in urbanization, and jumps by 25 percent for eastern counties. (The value of $a = 30.5$ is merely a computational constant without special meaning.) The X variables are usually interval, but dichotomous variables (often called "dummy variables" for some odd reason) can also be used. Ranked and categorical X variables cannot be used. Categorical variables may, however, be broken into several dichotomous variables. For example, if X_3 were a categorical variable with value 1 for east, 2 for west, 3 for south, and 4 for north, it could be broken into three dichotomous variables: $X_3 = 1$ for east and 0 otherwise, $X_4 = 1$ for west and 0 otherwise, and $X_5 = 1$ for south and 0 otherwise.

The b coefficients show how much Y changes with corresponding changes in the independent variables, but they do not directly show the relative importance of the independent variables in terms of influencing Y. This is shown by the "beta coefficients" (also printed out by library programs), whose formula is

$$\beta_i = b_i \frac{\text{(st. dev. of } X_i)}{\text{(st. dev. of } Y)}$$

The larger the value of β_i the more important is X_i in predicting the value of Y. The interpretation of the beta coefficients is fairly simple: an increase of X_i by 10 percent of the standard deviation of X_i will increase Y^* by $(.10)\beta_i$ standard deviations of Y. Unlike the b coefficients, the beta coefficients can be compared with each other and the importance of each independent variable stated exactly.

To compute the regression coefficients when $K = 2$ and $Y^* = a + b_1X_1 + b_2X_2$, let

$$P = \Sigma X_1^2 - \frac{(\Sigma X_1)^2}{N} \qquad Q = \Sigma X_2^2 - \frac{(\Sigma X_2)^2}{N}$$

$$S = \Sigma Y^2 - \frac{(\Sigma Y)^2}{N} \qquad T = \Sigma X_1X_2 - \frac{(\Sigma X_1)(\Sigma X_2)}{N}$$

$$V = \Sigma X_1Y - \frac{(\Sigma X_1)(\Sigma Y)}{N}$$

and

$$W = \Sigma X_2Y - \frac{(\Sigma X_2)(\Sigma Y)}{N}$$

Then,

$$b_1 = \frac{QV - TW}{PQ - T^2} \qquad b_2 = \frac{PW - TV}{PQ - T^2}$$

$$a = \frac{1}{N}(\Sigma Y - b_1\Sigma X_1 - b_2\Sigma X_2)$$

and

$$R^2 = \frac{b_1V + b_2W}{S} \text{ qed.}$$

For hand computation, therefore, it is necessary to compute from the original data the values of ΣX_1, ΣX_2, ΣY, ΣX_1^2, ΣX_2^2, ΣY^2, $\Sigma X_1 X_2$, ΣX_1Y, ΣX_2Y, where summation, of course, is over all N subjects, and insert the obtained values in the formulas above. The beta coefficients can then be calculated by observing that $P = N(\text{st. dev. of } X_1)^2$, $Q = N(\text{st. dev. of } X_2)^2$, and $S = N(\text{st. dev. of } Y)^2$.

The most important pitfall in multiple regression analysis comes when "multicollinearity" is present, that is, when two or more of the independent variables happen to be highly correlated with each other. When the r between X_i and X_j is greater than about $+.7$ (or less than $-.7$), it is impossible to distinguish between their effects on Y, so all the partial regression coefficients can be misleading. The remedy is simply to check the matrix of correlations among the X variables to see if any two are in fact highly correlated. One of the X variables can be discarded to remove the high correlations, and the analysis can proceed with much gained and little lost. If a certain R^2 is obtained by using a set of X variables, a higher R^2 will always result from adding one more X variable. The increase in R^2 shows how much extra explanatory power the added variable has after the other independent variables have explained their share. One other note of caution is in order. When the N subjects are in fact years, and the variables are time series, then very difficult problems typically arise. Their solution involves advanced procedures that are covered in textbooks of econometrics.

Significance

The significance of the association between variables is the confidence the researcher has that the calculated strength and form accurately reflect the true strength and form of the relationship between attributes of people, institutions, or events. If in reality attributes X and Y are unrelated, their true strength of association is near zero and their form of relationship is unimportant. But if in reality X and Y are closely related, the strength of association is high and the form is a significant topic for further research. However, the *calculated* strength and form may differ from the true strength and form because of five sources of distortion: data unreliability, form misidentification, sampling error, uncontrolled variables, and the ecological fallacy.

First, the data used in computation may be unreliable—that is, contaminated with errors of observation, measurement, or transcription.[5] If the errors in X are independent of those in Y (that is, the errors are uncorrelated), then the calculated coefficient of correlation between X and Y will be closer to zero than the true coefficient based on error-free data. Usually the depressing effect of random error is relatively small. If the errors are large in proportion to the true values of X and Y, and if X errors are highly correlated with Y errors (which can happen if all the data comes from a single unreliable source), then the calculated correlation will be higher (closer to $+1.0$ or -1.0) than the true correlation.

Second, the true form of association may be difficult to identify. If the true form is nonlinear, measures of association based on linear forms (Pearson's r) or monotonic forms (Spearman's rho) will understate the true strength of association. For example the scatter diagram in Figure 3.4a shows a hypothetical joint distribution of wealth (the independent variable) and political radicalism for $N = 19$ subjects; the rich and poor are less radical than those in the middle. The Pearson r will be close to zero since a linear form (Figure 3.4b) is little help in estimating radicalism from wealth. A nonlinear regression form, however, will produce a curve that fits the observed data fairly well (Figure 3.4c). However, the best solution may involve the categorization of the subjects into low, middle, and high wealth classes; inside each class there may be no relationship between wealth and radicalism (Figure 3.4d).

Sometimes the scatter diagram is so ambiguous that several alternative forms of association appear equally likely and the danger of misidentification becomes acute. Figure 3.5a shows an example, along with linear (Figure 3.5b) and curvilinear (Figure 3.5c) regression lines. Figure 3.5d shows the results of categorizing the observations into three classes and drawing regression lines for each. Mathematically the single linear regression line is the simplest, but the other forms give rise to very different explanations of the pattern of the scatter diagram, as can be seen if one tries to estimate the value of Y^* for an X to the right or left of the plotted values of X.[6]

The third type of problem arises when the correlations are based on data from a sample of subjects, not from the total population which is the target for the analysis.[7] Even if the sample is drawn by means of careful random methods an element of uncertainty exists: the sample coefficient may be larger or smaller than the population coefficient. The larger the size N of the sample, the smaller will be the likelihood of great deviations from the population coefficient. The nomograph in Figure 3.6 shows the confidence belts for Pearson r coefficients corresponding to samples of size $N = 3$, 4, 5, 7, 8, 10, 12, 15, 20, 25, 50, 100, 200, and 400. This is the nomograph for the 99 percent confidence level, which means that only one sample in every 100 will produce a correlation coefficient outside the belts shown. The nomograph is simple to use. Suppose that one calculates $r = +0.60$ for a sample of size $N = 25$. Enter the graph along the top and bottom at the vertical line $+.6$; go up and down the line spotting the two points where it intersects curved lines marked 25. Read across to the numbers along the right hand margin, which in this case are $+.13$ and $+.84$. With probability of 99 percent (that is, 99 to 1 odds) the true population coefficient lies between $+.13$ and $+.84$. Only one sample in 100 of size 25 will give an $r = .6$ when the population coefficient is above $+.84$ or below $+.13$.

The 99 percent confidence belts in the nomograph require a large sample before the investigator can be confident that the sample r is close to the population r. However, the nomograph can easily be used to obtain the 80 percent and the 90 percent confidence belts. To get the 80 percent belts, follow the same procedure except that instead of finding the intersection of $r = +.6$ with $N = 25$, find the approximate intersection with the bands $N^* = 4N - 9$. For sample size $N = 25$ and $r = +.6$, the bands are $N^* = 4(25) - 9 = 91$, or just outside the bands marked 100. This gives the confidence limits of $+.4$ and $+.75$. Thus only one sample in five (20 percent) of size 25 will give an r of $+.6$ when the population r is greater than $+.75$ or less than $+.4$. To get the 90 percent confidence belts follow the same procedure with $N^* = 2.5N - 4$; in this case $N^* = 2.5(25) - 4 = 58$, and the limits are $+.3$ and $+.77$ for the population r. The nomograph can be used without change for finding the 90 percent confidence belts for tetrachoric r.

The fourth condition affecting significance is the possibility that a third (or a fourth, and fifth) variable Z may be affecting both X and Y. It is possible that Z tends to lower X and Y for the same subjects, and to raise X and Y for other subjects, thus producing a higher coefficient of correlation between X and Y. Occasionally variable Z may raise X and lower Y for some subjects, and have the opposite effect for others, thus depressing the correlation coefficient. If there is serious concern about the possible effects of a third variable, a simple procedure is available that statistically removes all the (linear) effects of Z on

Figure 3.4a.   Ease of misidentification of form of relation.

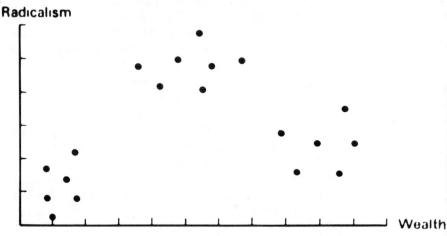

Radicalism

Wealth

Figure 3.4b,c,d.   Ease of misidentification of form of relation

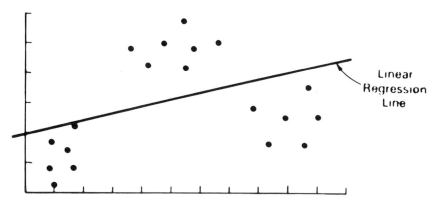

Linear
Regression
Line

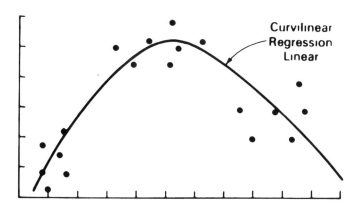

Curvilinear
Regression
Linear

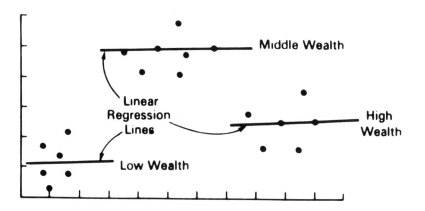

Middle Wealth

Linear
Regression
Lines

High
Wealth

Low Wealth

the relationship between X and Y. This technique is "partial correlation." The partial coefficient of correlation between X and Y, controlling for Z, is written $r_{xy \cdot z}$. The correlation between x and y not controlling for z is called the simple correlation, and is written r_{xy}. The simple correlation has already been discussed, and it is a welcome fact that the partial correlations $r_{xy \cdot z}$, $r_{xz \cdot y}$, and $r_{yz \cdot x}$ can be computed directly from the three simple correlations r_{xy}, r_{xz}, and r_{yz} by the formula:

$$r_{xy \cdot z} = \frac{r_{xy} - (r_{xz})(r_{yz})}{\sqrt{(1 - r_{xz}^2)(1 - r_{yz}^2)}}$$

Note that $r_{xy} = r_{yx}$ and that $r_{xy \cdot z} = r_{yx \cdot z}$, but that $r_{xy \cdot z} \neq r_{xz \cdot y}$. Computation of the denominator in the above expression is simplified somewhat by the table of values of $\sqrt{(1 - r^2)}$ in the Appendix. For example, if $r_{xy} = +.8$, $r_{xz} = +.7$, and $r_{yz} = +.9$, then

$$r_{xy \cdot z} = \frac{.8 - (.7)(.9)}{\sqrt{(1 - (.7)^2)(1 - (.9)^2)}} = \frac{.8 - .63}{\sqrt{(1 - .49)(1 - .81)}}$$

$$= \frac{.17}{\sqrt{(.51)(.19)}} = \frac{.17}{\sqrt{.097}} = \frac{.17}{.31} = .55$$

Thus the partial correlation between X and Y controlling for Z is $+.55$, somewhat less than the simple correlation of $+.8$. Z has acted to raise the correlation between X and Y.

Similarly we have

$$r_{xz \cdot y} = \frac{.7 - (.9)(.8)}{\sqrt{(1 - .9^2)(1 - .8^2)}} = \frac{.70 - .72}{\sqrt{(1 - .81)(1 - .64)}}$$

$$= \frac{-.02}{\sqrt{(.19)(.36)}} = \frac{-.02}{.26} = -.08$$

Thus the partial correlation between X and Z controlling for Y is a *negative* .08, while the simple correlation was a high positive .7. Y has greatly inflated the strength of association between X and Z.

Thus a third variable, sometimes called a "control variable," can raise or lower a simple correlation, or leave it unchanged. Note that if $r_{xy} = r_{xz} r_{yz}$ then the partial coefficient $r_{xy \cdot z} = 0$. The formula for partial coefficients can also be used for Spearman rho. The field of "causal analysis" developed by Herbert Simon is concerned with chains of causation and intercorrelation when there are three or more variables involved. To be a true control variable Z must be a logically or temporally prior attribute—it must somehow come before X and Y so that it can jointly influence those two attributes. Unfortunately, causal analy-

Figure 3.5a.

Figure 3.5d.

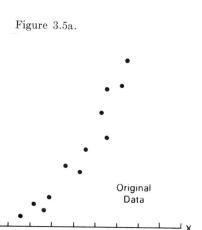

Original
Data

Figure 3.5b.

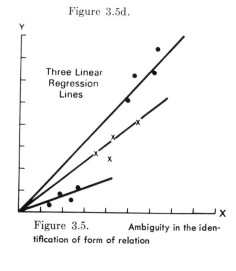

Three Linear
Regression
Lines

Figure 3.5.   Ambiguity in the identification of form of relation

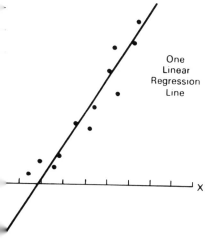

One
Linear
Regression
Line

Figure 3.5c.

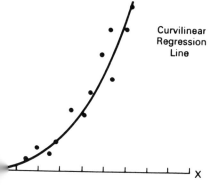

Curvilinear
Regression
Line

sis has thus far paid little attention to the problem of temporal priority of variables.[8]

In the case of dichotomous variables a control can be introduced by splitting the original 2×2 table into two such tables, one for those subjects who score 'yes' on Z and the other for those who score 'no.' Theoretically almost anything can happen to the original ϕ correlation, as the following examples show.

		SIMPLE CORRELATION		PARTIAL CORRELATION				
				$Z = 1$		$Z = 0$		
		$X = 1$	$X = 0$	$X = 1$	$X = 0$	$X = 1$	$X = 0$	
Ex. 1	$Y = 1$	10	4	5	2	5	2	Pattern
	$Y = 0$	4	10	2	5	2	5	unchanged
		$\phi = .43$		$\phi = .43$		$\phi = .43$		
Ex. 2	$Y = 1$	52	20	2	10	50	10	Pattern
	$Y = 0$	20	52	10	50	10	2	vanishes!
		$\phi = .44$		$\phi = 0$		$\phi = 0$		
Ex. 3	$Y = 1$	10	10	8	2	2	8	Opposing
	$Y = 0$	10	10	2	8	8	2	patterns emerge!
		$\phi = 0$		$\phi = +.6$		$\phi = -.6$		
Ex. 4	$Y = 1$	10	10	5	9	5	1	Similar
	$Y = 0$	10	10	1	5	9	5	patterns emerge!
		$\phi = 0$		$\phi = +.19$		$\phi = +.19$		

The search for control variables that may be producing or suppressing a high correlation among variables is a central theme of the scientific method. It lies at the heart of the concept of the controlled experiment in the physical and biological sciences. Scholars working from documents and other nonexperimental data can sometimes approximate controlled experiments by comparing two situations that seem to be identical except for one or two variables. Such situations are not often encountered. More often one is faced with a number of situations and several variables each of which varies in the different situations. Multivariate and causal analysis can make sense out of these situations only if the investigator has a theoretical model to begin with. He can feed the data into the model and see how close the expected outcome (the value of the dependent variable) is to the observed outcome. If the fit is good, the model works; otherwise it must be modified. The most common types of multivariate analysis are factor analysis (which reduces a large number of intercorrelated variables to a handful) and multiple regression analysis, which builds estimating equations for one dependent variable from a number of independent variables.

Fifth, the significance of association may be confused by the blurring of the distinction between individual and ecological attributes. Individual attributes such as age, sex, church affiliation, voting record, and wealth have meaning only for individual people. Ecological attributes are attributes of geographical units, such as area, population, metropolitan-hinterland status, control by oligarchy, volume and direction of trade, population growth rate, and form of government. Some ecological attributes are average values of individual attributes for the individuals who live in the area, such as Democratic share of the vote, per capita income, median age, percent unemployed, and proportion of Catholics. Yet these variables describe areas and communities, not individuals. The ecological fallacy is committed by drawing conclusions about individual patterns from ecological data, and can take a variety of forms. A lesser fallacy, the "individualistic fallacy" deduces ecological attributes from individual data—such as the conclusion that a nation is aggressive in international affairs on the evidence that the average citizen has aggressive drives.[9]

Given the availability of ecological data from governmental sources, and the desirability of analyzing individual behavior, it is easy to commit the fallacy, especially in voting research. If the correlation between percent Catholic and percent Democratic for a number of counties is high, the "obvious" but fallacious conclusion is that Catholics voted Democratic. They may have—or perhaps they did not, as the following extended example demonstrates.

Suppose that a certain area contains 2000 Catholic and 3000 non-Catholic voters. Suppose also that 60 percent of the Catholics and 40 percent of the others voted Democratic in a particular election. An interview survey of all 5000 people would produce the 2×2 table shown as Table 3.15. Suppose further that the population lives in five precincts of equal size, with Catholics comprising 20 percent, 30 percent, 40 percent, 50 percent, and 60 percent respectively of the voters. Finally (the only unrealistic assumption) suppose that 60 percent of the Catholics and 40 percent of the others in *each* precinct vote Democratic. Table 3.16 shows the precinct breakdown; note that $\phi = +.2$ for each precinct.

Now make the very realistic assumption that the interview data on which the 2×2 tables were based do not exist—all that is available are the *marginal* totals and percentages. That is, the number and percentage of Catholics and non-Catholics, and Democrats and Republicans is known for each precinct, but the cross-tabulations are unknown. Simple inspection of precinct "C" shows that almost any number of Catho-

TABLE 3.15   Hypothetical Distribution of the Votes of
              Catholics and Non-Catholics

The ϕ coefficient between "Catholic" and "Democrat" is:

$$\phi = \frac{1200 \times 1800 - 1200 \times 800}{\sqrt{2000 \times 3000 \times 2400 \times 2600}} = +.196 \approx 60\% - 40\%$$

Note that the approximation to ϕ, the difference of proportions, equals $+.20$, very close to the actual value of $+.196$.

	CATH.	NON-CATH.	Σ	%
Dem.	1200	1200	2400	48%
GOP	800	1800	2600	52%
Σ	2000	3000	5000	100%
%	40%	60%	100%	

TABLE 3.16   Hypothetical Precinct Distributions of
              Voters

	CATHOLIC	NON-CATHOLIC	Σ	%
PRECINCT "A"				
Dem	120	320	440	44%
GOP	80	480	560	56%
Σ	200	800	1000	100%
%	20%	80%	100%	
PRECINCT "B"				
Dem	180	280	460	46%
GOP	120	420	540	54%
Σ	300	700	1000	100%
%	30%	70%	100%	
PRECINCT "C"				
Dem	240	240	480	48%
GOP	160	360	520	52%
Σ	400	600	1000	100%
%	40%	60%	100%	
PRECINCT "D"				
Dem	300	200	500	50%
GOP	200	300	500	50%
Σ	500	500	1000	100%
%	50%	50%	100%	
PRECINCT "E"				
Dem	360	160	520	52%
GOP	240	240	480	48%
Σ	600	400	1000	100%
%	60%	40%	100%	

lic-Democrats, from 0 to 400, is consistent with the observed marginal totals. The percentages can be cast in the form of Table 3.17, or in the scatter diagram of Figure 3.7.

TABLE 3.17 Observed Ecological Data for Five Precincts

PRECINCT	X = % CATH	Y = % DEM	N
"A"	20%	44%	1000
"B"	30%	46%	1000
"C"	40%	48%	1000
"D"	50%	50%	1000
"E"	60%	52%	1000
Σ	40%	48%	5000

When plotted in Figure 3.7, the five points fall exactly on a straight line, indicating that the Pearson $r = +1.0$. Thus the ecological correlation, $+1.0$, is much higher than the individual correlation $\phi \approx +.2$. Indeed for $\phi = +1.0$, *all* the Catholics would have to vote Democratic, and *all* the non-Catholics vote Republican! The ecological regression equation, $Y^* = .2X + .40$, provides a way to estimate exactly how a precinct will vote knowing what proportion of Catholics it contains. The ecological form explains the behavior of *precincts,* but not *voters.* To use the ecological correlation as an approximation of the individual correlation is totally without justification—it would be a classic example of the ecological fallacy.

The rediscovery of the ecological fallacy in 1950 had a disastrous impact on aggregate voting analysis. No longer could anyone maintain confidently that a high correlation between percent German and percent decline in Republican vote proves that Germans switched away from the GOP. The pessimism was not, however, entirely justified, for there are four escape routes from the dilemma of the fallacy. First, the entire analysis can be couched in individual terms using interview data. This method is very fruitful, but extraordinarily expensive (a nationwide survey of 1000 people may cost $25,000 or more), and strictly limited in application. Second, the entire analysis can be put in ecological terms—if Republican losses correlate at $+.95$ with percent German, one *can* say that the GOP lost most in geographical areas where there are many Germans, and lost least elsewhere. This approach suggests that the Germans *or their neighbors* were responsible for the decline, and clearly points the direction for further analysis. A third approach is to locate homogeneous areas (for example, precincts that are more than 50 percent German) and study the aggregate patterns there. Unfortunately, many interesting groups (businessmen, British immigrants) may not constitute a majority in any precinct. Furthermore, highly homogeneous areas are rather odd, and may not reflect the patterns of the whole group. There is reason to doubt that Germans who concentrate in all-German neighborhoods behave exactly the same as those who scatter throughout the community. But for geographically segregated groups, such as farmers, urban Negroes, small-town dwellers, and suburbanites, the third approach may prove quite satisfactory.

Figure 3.6.   99% Confidence belts for Pearson r

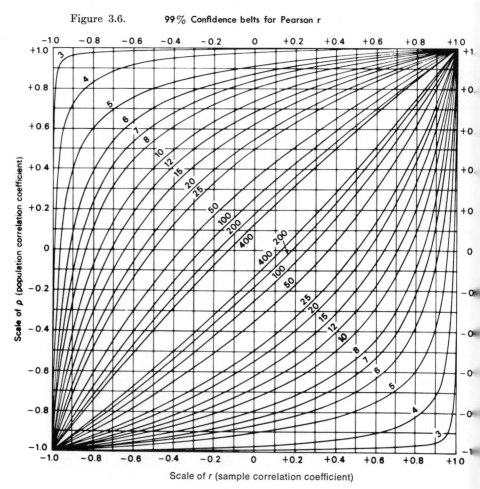

Scale of r (sample correlation coefficient)

The numbers on the curves indicate sample size. The chart can also be used to determine upper and lower 0.5% significance points for *r*, given *p*.

Source: Biometrika Tables (1965)

The fourth approach involves the estimation of ϕ and the 2×2 cell entries directly from the aggregate data, but it is not highly reliable. This method requires that the ecological r be rather high (at least $+.7$), and makes the strong assumption that members of each group behave the same no matter who their immediate neighbors are. Thus in the

Figure 3.7.   Scatter diagram for data in table 3.17

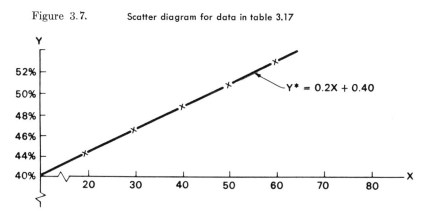

(The broken lines on the X and Y axes indicate that only the relevant portion of the whole scatter diagram is shown.)

example it assumes that Catholics in each precinct vote Democratic at the same rate, and likewise the non-Catholics. If these conditions are met, then $\phi \approx b$, where b is the ecological regression coefficient. More precisely:

$$\phi \approx b \sqrt{\frac{X(1-X)}{Y(1-Y)}} = b \sqrt{\frac{(NX)[N(1-X)]}{(NY)[N(1-Y)]}}$$

where X is the mean proportion of the independent variable in the entire population (here X = percent Catholic = 40 percent), and Y is the mean of the dependent variable (here Y = percent Democratic = 48 percent). The values NX, $N(1-X)$, NY, and $N(1-Y)$ represent the total number of Catholics, non-Catholics, Democratic votes, and Republican votes. If *both* the X and Y percentages for the entire population lie between 25 and 75 percent, then $\sqrt{[(X)(1-X)]/[(Y)(1-Y)]}$ lies between 0.85 and 1.15, so that $\phi \approx b$ with an error of less than 15 percent. Note that if some Catholics and non-Catholics do not vote, the additional assumption is necessary that the nonvoters are divided religiously the same as the total population. Another hidden assumption, that all precincts have the same population, is usually not important. The Pearson coefficients can be computed when the units (say, counties or states) have greatly different populations. Empirically, however, the "weighted" coefficients are usually quite close to the ordinary "unweighted" kind.

The regression equation for the scatter diagram in Figure 3.7 is $Y^* = .2X + .40$, hence $b = .2$ and $\phi \approx b = .2$ is a very good estimate. The justification for the fourth method is not difficult. The coefficient b equals the *increase* in Y^* when X increases one unit. Thus if X is increased from 40 to 50 percent, Y^* increases by $10b = 2$ percent. That is, for every 100 voters there are now 50 Catholics instead of 40, and 50 non-Catholics instead of 60. If 60 percent of all Catholics and 40 percent of the others vote Democratic, a shift from 40–60 to 50–50 means a net increase of 2 Democrats (that is, six new Catholic Democrats minus four old non-Catholic Democrats). Thus b equals the difference between the percentage of Catholics who are Democrats (60 percent) and the percentage of non-Catholics who are Democrats (40 percent). In terms of the 2×2 table in Table 3.18:

$$
\begin{aligned}
b &= \frac{p_{11}}{p_{11} + p_{21}} - \frac{p_{12}}{p_{12} + p_{22}} \\
&= \frac{p_{11}p_{12} + p_{11}p_{22} - p_{11}p_{12} - p_{21}p_{12}}{(p_{11} + p_{21})(p_{12} + p_{22})} \\
&= \frac{p_{11}p_{22} - p_{21}p_{12}}{(p_{-1})(p_{-2})}
\end{aligned}
$$

However

$$
\begin{aligned}
\phi &= \frac{p_{11}p_{22} - p_{21}p_{12}}{\sqrt{p_{-1}p_{-2}p_{1-}p_{2-}}} = \frac{(p_{11}p_{22} - p_{21}p_{12})}{p_{-1}p_{-2}} \sqrt{\frac{p_{-1}p_{-2}}{p_{1-}p_{2-}}} \\
&= b \sqrt{\frac{p_{-1}p_{-2}}{p_{1-}p_{2-}}} \qquad \text{qed, since } p_{-1} = X, \ p_{-2} = 1 - X \\
&\qquad\qquad\quad \text{and} \quad p_{1-} = Y, \ p_{2-} = 1 - Y
\end{aligned}
$$

The actual cell entries in Table 3.15 or Table 3.18 can be estimated directly from the regression coefficients and the aggregate totals. A little algebra is enough to prove the following formula for the value of p_{11}: $p_{11} = XY + bX(1 - X)$ and $Np_{11} = (NX)[Y + b(1 - X)]$ where X = proportion of Catholics (the independent variable) $= .40$, and Y = proportion of Democrats in the population (the dependent variable) $= .48$. Thus $p_{11} = (.40)(.48) + (.20)(.40)(.60) = .192 + .048 = .24$, and $Np_{11} =$ the number of Catholic Democrats $= (2000)(.48 + .2 \times .6) = .6 \times 2000 = (.24)(5000) = 1200$, and the other cell entries can be found by subtraction from the marginal values, qed.

In most research, especially the analysis of voting statistics, the assumption that the members of each group behave the same no matter where they live is unrealistic. Ecological (neighborhood) differences may be of considerable importance. For example, precinct A may be an affluent suburb, B an urban white collar area, C the home of a powerful political organization, D a rapidly changing neighborhood, and E a heavily unionized blue collar area. Or there may be "interaction effects" present. That is, Catholics may vote differently in precinct A where they are

TABLE 3.18 Percentage Distribution, $N = 1.00$

	CATH	NON-CATH	Σ
Dem	p_{11}	p_{12}	p_{1-}
GOP	p_{21}	p_{22}	p_{2-}
Σ	p_{-1}	p_{-2}	1.0

a small minority than in precinct E where they are a majority because of the influence of the predominantly Republican non-Catholics. Furthermore the composition of the Catholic and non-Catholic population may vary widely—Irish Catholics in A, Germans in C, Poles in E; old-stock Congregationalists in A, second generation Jews in C, and Negro Baptists in D. It can hardly be expected that "Catholics" and "non-Catholics" have the same attitudes and the same voting patterns in each area. Unless more precincts are studied, more variables added, and advanced multivariate methods used, significant conclusions will be hard to reach. Note that the same kinds of interaction and ecological effects complicate analysis of survey data, but they are virtually impossible to handle in national surveys.

The significance of correlations for the research project lies in their suggestiveness for further research, especially more precise definition of groups and attributes. The best pattern-searching strategy is to compute the matrix of correlations between every pair of variables and search for clusters of variables that are all highly correlated with each other. The construction of the matrix may be expensive, since there will be $\frac{1}{2}K(K-1)$ correlations for K variables. That is, three correlations for three variables, 6 for 4, 10 for 5, 28 for 8, 45 for 10, 190 for 20, and 4750 correlations for 100 variables. A variable that does not correlate at $r = +.5$ or higher with any other can be dropped, thus reducing the number of variables and simplifying the matrix. Factor analysis (done on a computer) can be used to reduce a matrix to a small number of dimensions, but the technique is very complicated. Much simpler are the methods of cluster analysis described in Chapters 4 and 6. If the subjects are countries, a matrix of correlations may reveal one cluster of attributes consisting of per capita GNP, literacy, urbaniza-tion, newspaper readership, and low birth rates; a second cluster may include strength of labor movement, level of freedom of the press, and degree of political party competition; another cluster might include de-gree of social tension, extent of political stability, and political strength of the military. The first cluster can be called "development," the second "democracy," and the third "stability." The clusters will resemble the dimensions that a factor analysis would extract.

The techniques described in the next chapter for graphing the clus-tering patterns of agreement scores can easily be adapted to the problem of graphing clusters of correlated attributes. Once clusters have been found and identified, the challenge of analyzing, understanding, and ex-plaining the complexities of social structure and behavior has just be-gun—but it is off to a flying start.

FOOTNOTES

1. This paragraph is a bit technical and may be skipped.
2. See Theodore A. Anderson and Morris Zelditch, *A Basic Course in Statistics* (New York: 1968), pp. 146–57, for the computation by hand of gamma and related measures known as tau and Somers' d.
3. Multiple regression analysis is the basic technique in econometrics. The best introductory text is J. Johnston, *Econometric Methods* (New York: 1963), which introduces the matrix algebra needed to handle a lot of equations concisely and elegantly. Karl A. Fox, *Intermediate Economic Statistics* (New York: 1968) is the most useful book on econometric analysis of actual data.
4. See Donald J. Veldman, *Fortran Programming for the Behavioral Sciences* (New York: 1967), pp. 294–307 for the write-up of a typical multiple regression program.
5. See Paul M. Siegel and Robert W. Hodge, "A Causal Approach to the Study of Measurement Error," in Hubert and Ann Blalock eds., *Methodology in Social Research* (New York: 1968), pp. 28–59. Helen Walker and Joseph Lev, *Statistical Inference* (New York: 1953), pp. 289–314.
6. The straightforward but time-consuming techniques for fitting cur-vilinear regression forms can be found in Frederick Croxton, Dudley Cowden, and Sidney Klein, *Applied General Statistics* (Englewood Cliffs: 1967) Ch. 20. For more detail see Mordecai Ezekiel and Karl Fox, *Methods of Correlation and Regression Analysis* (New York: 1959). On the problem of identification see Robert Fogel, 'The Speci-fication Problem in Economic History," *Journal of Economic History,*

27 (1967), pp. 283–308; for an advanced and comprehensive treatment, see Franklin M. Fisher, *The Identification Program in Econometrics* (New York: 1966).

7. The discussion of sampling error in Chapter 1, although concerned with only one variable, is relevant to this paragraph.

8. For an introduction to causal analysis see Hayward Alker, *Mathematics and Politics* (New York: 1965) Ch. 6, and Hubert Blalock, *Causal Inferences in Non-experimental Research* (Chapel Hill, N.C.: 1964). For advanced discussion see Blalock and Blalock, *Methodology in Social Research,* pp. 155–235.

9. On the ecological fallacy generally, see Mattei Doggan and Stein Rokkan, eds., *Quantitative Ecological Research.* (Cambridge: 1969), which has guides to the literature. For further bibliography, see Chapter **7** below.

SPECIAL PURPOSE
STATISTICS

CHAPTER FOUR   The research design determines
which specific statistical tools
should be used in analyzing his-
torical data. Chapters 2 and 3 reviewed techniques that work well with
most research designs. This chapter presents several methods for handling special problems in historical analysis. Included are roll-call analysis
techniques, the agreement index and the Guttman scale, along with the
clustering dendrogram (adapted from zoological taxonomy) and the
success score. This chapter also reviews the standard statistical methods
for measuring time series, trends, growth rates, and inequality and
segregation.

Legislative Roll-Call and Clustering Techniques

The historian studying roll call votes of legislatures, judicial bodies,
conventions or committees needs to identify clusters of men who constituted voting blocs and the clusters of roll-calls or issues in which the
blocs appeared, and if possible, he would like to discover underlying
attitudes the legislators expressed through their votes.

Consider the case of a legislature in which N men can respond
'yea,' 'nay,' or 'absent' on R roll calls. The total number of responses
$N \times R$ is too large to rely upon unsystematic impressions. The first
question to ask is what is the underlying pattern of agreement between
legislators. Considering the extensive amount of work a full analysis
of agreement patterns entails, a computer is usually necessary. For any
pair of legislators A and B, there are nine possible patterns of response
to a single roll call vote, as shown in Table 4.1. Each of the R roll

calls may be assigned to one of the nine cells. Thus A voted yea on $a + b + c$ roll calls; of these B voted yea a times, nay c times, and was absent b times. B voted nay $c + f + i$ times; of these A voted yea c times, nay i times, and was absent f times.

TABLE 4.1   Distribution of Joint Responses of Two Legislators to R Roll Calls.

| | | LEGISLATOR B | | | |
		YEA	ABSENT	NAY	Σ
	yea	a	b	c	$a + b + c$
Legislator A	abst	d	e	f	$d + e + f$
	nay	g	h	i	$g + h + i$
	Σ	$a + d + g$	$b + e + h$	$c + f + i$	R

"Agreement" can be defined in several ways. We can restrict the definition of agreement to those cases in which both men voted yea, cell a, or both voted nay, cell i; disagreement can be defined either as the remaining seven cells, or more simply the number of times the men voted against each other (cells c and g). Thus the historian can use either: Agreement of A and B =

$$A^*(A,B) = \frac{a + i}{a + c + g + i}$$

or

$$A(A,B) = \frac{a + i - c - g}{R}$$

The latter form has several desirable properties: for every pair of legislators the denominator is the same; agreement can range from $+1.0$ (both men always vote, and always vote together), to -1.0 (both men always vote, and vote against each other); if one man is absent, the second index counts it as half agreement and half disagreement; and the index is zero when $a + i = c + g$.

"Cohesion" is the behavior of a subgroup of members on one roll call. If a subgroup (say, the Republicans) votes together, it displays perfect cohesion. If the subgroup is evenly divided between ayes and nays, it displays zero cohesion. Thus we define the "Rice index of cohesion" as

$$C = \frac{\text{yeas} - \text{nays}}{\text{membership}} \quad \text{or} \quad \frac{\text{nays} - \text{yeas}}{\text{membership}}$$

whichever is positive. The average cohesion for a subgroup in a legislature is simply its mean cohesion score on all R roll calls.

"Likeness" is the similarity in voting patterns of two subgroups on one roll call. If subgroup I responded with a yeas, b absences, and c nays, while subgroup II responded to the roll call with d yeas, e absences, and f nays, the Index of Likeness equals

$$L = 1 - \left(\frac{a}{a + c} - \frac{d}{d + f} \right) = 1 - \frac{a}{a + c} + \frac{d}{d + f}$$

From time to time use has been made of correlation coefficients in roll call analysis. Examples were given in Chapter 3 that involved the prediction of votes from characteristics of the legislators. There seems to be little reason to compute ϕ coefficients for pairs of roll calls or pairs of legislators. However, the Q coefficient has an important use in the construction of Guttman scales.

The cohesion of various groups of legislators (Republicans, Southerners, rural constituencies, leaders, and so on) can be precisely described by the index of cohesion. The higher the index, the more the group held together, which may suggest an important role for organization, leadership, or common interests of one sort or another. If a group shows low cohesion, one can conclude its leadership was mediocre, or that the members in fact had little in common.

Agreement scores can be used to identify the actual blocs of men who tended to vote together on all or on selected issues. To find the blocs, compute the agreement scores for each of the $\frac{1}{2} N(N - 1)$ pairs of legislators. Then search for *clusters* of like-voting men. The clusters consist of K men whose $\frac{1}{2} K(K - 1)$ agreement scores with each other exceed a specified level, say $+.50$. Alternatively, one can require that the mean agreement score in a cluster exceed $+.50$, and that no single agreement score be below $+.40$. The formation of clusters is not an automatic process. The minimum agreement scores can vary depending on the circumstances. If a low minimum is chosen, probably a few large clusters will emerge, with nearly every legislator included in one, and possibly some men included in two different clusters. If a very high minimum is chosen, a few small clusters will emerge, and probably most of the legislators will be excluded from any of them. Sometimes the historian has additional information about the legislators, and may wish to use it to decide who should and should not be included in a cluster.

An interesting possible use of agreement scores based on dichotomous variables that need not be roll call votes is the identification of taxonomic groups. Suppose that N men have dichotomous (yes or no) scores on R attributes (for example, leader–nonleader, churchmember–nonchurchmember, local–cosmopolitan, popular–obscure, and so on). An agreement score can be defined for each pair of men as:

$$A(A,B) = \frac{a + d - b - c}{a + b + c + d} = \frac{a + d - b - c}{R}$$

$$= \frac{2(a + d)}{R} - 1 = 1 - \frac{2(b + c)}{R}$$

where a is the number of attributes for which both A and B answer yes, and d is the number of attributes for which both men answer no, and $b + c$ the number for which they differ. A matrix of agreement scores showing the value of A for every pair of men can be constructed, and clusters can be formed. In fact, "taxonomic trees" can also be formed by the following process. Search the matrix of A values for the pairs that agree at the $+.95$ level and cluster together all such men. Repeat the process at the $+.90$ level, this time permitting individuals or even clusters to merge with established clusters. Continue the process for $A = +.85, +.80, \ldots +.05$. Various rules for merging clusters have been suggested. One could require that every candidate for a cluster must agree with *every* member of the cluster at least at the stated level of A, or that the *mean* agreement of the candidate with the members be at the stated level, perhaps with the requirement of a certain minimum. The agreement matrix in Table 4.2 thus gives rise to the taxonomic trees (called "dendrogram" in biology) in Figures 4.1 and 4.2.

TABLE 4.2 Agreement Matrix for 10 Legislators on 6 Roll Calls

B	C	D	E	F	G	H	I	J	
				LEGISLATOR					
$-.33$	$.33$	$.00$	$.17$	$.33$	1.00	$.33$	$.33$	$.67$	A
1.00	$.33$	$.67$	$-.17$	$-.33$	$-.33$	-1.00	$-.33$	$.00$	B
	1.00	$.67$	$.67$	$-.17$	$-.33$	$.33$	$-.33$	$.33$	$.67$	C
		1.00	$-.50$	$-.67$	$.00$	$-.67$	$.00$	$.33$	D
			1.00	$.17$	$.17$	$.17$	$-.50$	$-.17$	E
				1.00	$.33$	$.33$	$.33$	$.00$	F
					1.00	$.33$	$.33$	$.67$	G
						1.00	$.33$	$.00$	H
							1.00	$.67$	I
								1.00	J

The agreement matrix in Table 4.2 was computed from the six hypothetical roll calls given in Table 4.6. Since the matrix is symmetrical (that is, $A(A,B) = A(B,A)$) only the upper half is shown.

The construction of a "tree" is based on the rule that a candidate for a cluster must have the minimum agreement score indicated with each of the members of the cluster. To construct it, we begin with the pair of men *A* and *G*, whose agreement score is +1.00. They cluster at the *A* = 1.00 level as shown. Next we notice that *J* has an agreement score of +.67 with both A and G; although *J* has an agreement of +.67 with *C* and in the absence of other information, it is best to cluster it with the largest group, in this case with cluster *A–G*. Then we note

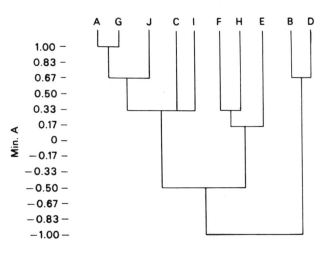

Figure 4.1.   Taxonomic tree (dendrogram) based on Table 4.2

Figure 4.2.   Alternative taxonomic tree based on Table 4.2

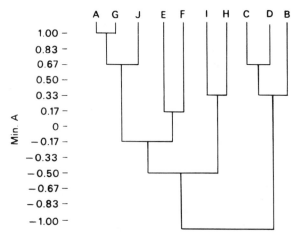

that C and I cluster with A–G–J at the $+.33$ level, since .33 is the minimum agreement score among the five men A, G, J, C, and I. This five man cluster does not join any other cluster above the .00 level, so we turn to the other seven men to search out clusters. We see that B and D cluster at the $+.67$ level, and form no new clusters. Perhaps we could have clustered C and D at the .67 level, with B joining at the .33 level, but then C would have to be removed from the larger cluster A–G–J–C–I. In deciding on assignments like this, there are often options, (compare Figure 4.2), and the historian should try various combinations. Finally, F and H cluster at .33, and E joins the cluster at .17. The tree is complete when all the branches finally join together at $A = -1.00$.

Power and Success Scores

The analysis of roll calls can be considerably improved by dropping the categories "yea-absent-nay" in favor of "win-absent-lose." Since the purpose of most roll calls is to arrive at a decision, it makes a great deal more difference who won and who lost, than who voted yea and nay. There will seldom be difficulty deciding who won a roll call—it could be a minority if a two thirds or three quarters majority was necessary for passage. Sometimes the legislature quickly reverses itself, or otherwise nullifies a roll call vote. The historian can decide for himself whether a nullified vote should be used in his analysis.

Once the original yea–nay record is transcribed into win–lose format, agreement and cohesion indices can be calculated from the same formulas, with win and lose substituted for yea and nay. The greatest advantage in using win–lose patterns is that the political outcome of the legislative process is kept clearly in mind at all times.

Perhaps the most useful legislative index is the "success score," which can be tabulated easily from the win–lose record. Out of R roll calls, A wins a times, loses c times, and was absent b times. His success score can be defined as

$$\text{Success Score of } A = SS(A) = \frac{a - c}{a + b + c + d}$$

$$= \frac{a - c}{R}$$

The success score ranges from $+1.0$ (when $a = R$) to -1.0 (when $c = R$), and equals zero when $a = c$. Absences leave the numerator unchanged, but an increase in the denominator pulls the success score closer to zero. A few examples show what the success score looks like. Table 4.3 gives the success score for a variety of combinations of wins, losses, and absences. In each case $R = 100$.

TABLE 4.3   Sample Success Scores

WINS = a	ABSENCES = b	LOSSES = c	SS
100	0	0	1.00
95	5	0	.95
95	0	5	.90
90	10	0	.90
70	20	10	.60
80	0	20	.60
50	50	0	.50
75	0	25	.50
50	30	20	.30
40	20	40	.00
50	0	50	.00
10	80	10	.00
0	90	10	−.10
45	0	55	−.10
20	50	30	−.10
10	70	20	−.10
30	30	40	−.10

The success score for a subgroup of N_i members is simply the mean SS for the members of the group, that is, the mean $SS = \dfrac{\Sigma SS_i}{N_i}$. The simple unweighted mean can be used since the denominator in each SS was the same number R. If we had defined the success score as wins divided by wins plus losses, the denominators might have varied widely, and the subgroup SS would not be easy to compute. Often newspapers or other sources routinely report the party breakdown of a roll call. Computation of each party's average SS for R roll calls then becomes much easier than computing the SS for any one member. The party SS is

$$\frac{\Sigma W_i - \Sigma L_i}{N_i \times R}$$

where W_i and L_i equal the number of party men who were on the winning and losing side of roll call $\#_i$, and N_i is the strength of the party in number of seats held. To prevent the denominator from changing, one should limit the computation of a party SS to one or two year periods.

The concept of success can be extended beyond roll calls; any type of election or voting process will do. Table 4.4 shows the success scores for 533 midwestern counties in the elections of 1888, 1890, 1892, 1894, and 1896, with the counties grouped into 16 categories based on level of urbanization and proportion of old-stock white voters. A county

TABLE 4.4   Mean Success Scores for 16 Categories of
Midwestern Counties, 1888–1896

(All counties in Illinois, Indiana, Iowa, Michigan, Ohio, and Wisconsin)

% URBAN IN 1900	% OLD-STOCK WHITE VOTERS IN 1910				Σ
	75–100%	50–75%	25–50%	0–25%	
Farm 0%	10	11	29	34	20
Rural 1–19.9%	3	8	17	40	11
Urban 20–49.9%	18	21	33	49	26
City 50–100%	28	26	39	65	42
Σ $N = 533$	12	16	29	47	23

achieved a perfect score (SS = 100) only if it gave its plurality five times for the winning party (the Democrats in 1890 and 1892, and the Republicans the other years). Note that each column represents counties with about the same proportion old-stock white voters. Reading down each column, the SS increases from rural (low) to farm, urban, and city. Thus SS increases with urbanization, holding the old-stock-white factor constant. Similarly each row represents counties with roughly the same level of urbanization. Reading across from left to right, for each row the SS increases as the proportion of old-stock whites declines. Furthermore, the two factors of urbanization and old-stock are additive—that is, the lowest scores are in rural old-stock counties (the upper left-hand four cells), and the highest scores are in the urban immigrant counties in the lower right-hand corner. The number of counties represented in each cell ranged from 5 (lower left corner) to 59 (top left corner), so the number of "votes" in each cell ranged from $5 \times 5 = 25$ to $5 \times 59 = 295$.

The concept of success can be further extended to cover the concept of 'power.' Power has proven remarkably difficult to define, but the best effort is that of Robert Dahl who defined the power of A over B, $P(A \text{ over } B)$, in terms of the likelihood that A would get his way when B opposed him versus the likelihood that B would get his way when A opposed him. Translated into legislative terms, the Dahl power index equals:

$$P(A \text{ over } B) = \frac{c - g}{a + c + g + i}$$

where the letters refer to the cell entries in Figure 4.3. That is, $P(A \text{ over } B)$ is the proportion of times when A wins and B loses, minus

Figure 4.3   Distribution of Joint Wins and Losses of Two Legislators on R Roll Calls

LEGISLATOR B

		WINS	ABSENT	LOSES	Σ
	wins	a	b	c	$a + b + c = W_A$
Legislator	absent	d	e	f	$d + e + f = A_A$
A	loses	g	h	i	$g + h + i = L_A$
	Σ	$a + d + g = W_B$	$b + e + h = A_B$	$c + f + i = L_B$	$a + b + c + d + e$ $f + g + h + i = R$

the proportion when B wins and A loses; the denominator is the number of times the two men both voted on a roll call. Unfortunately, the Dahl index of power is not transitive. Table 4.5 shows a hypothetical case in which $P(A \text{ over } B) = +1$, $P(B \text{ over } C) = +1$, and $P(C \text{ over } A) = +1$. That is, A is more powerful than B, B more than C, and

TABLE 4.5 Hypothetical Distribution of Wins and Losses on Six Roll Calls for a Subgroup of Three Legislators

$+$ = win; 0 = absent; $-$ = lose

ROLL CALL #	1	2	3	4	5	6
Mr. A	+	+	0	0	−	−
Mr. B	−	−	+	+	0	0
Mr. C	0	0	−	−	+	+

$$P(A \text{ over } B) = \frac{2-0}{2} = +1 = P(B \text{ over } C) = +1 = P(C \text{ over } A)$$

yet C is more powerful than A. The problems of Dahl index also extend to computational difficulty. To calculate all the power indices in a legislature of 50 men voting on 40 roll calls requires the construction of $\frac{1}{2}$ $50 \times 49 = 1225$ 3×3 tables like the one in Figure 4.3; to get the cell entries for each of these tables requires going through 40 votes by each man, or $(40 + 40) \times 1225 = 96,000$ operations in all!

Fortunately, we can devise a reasonable alternative power index that is transitive, and that can easily be computed for any pair of men without having to set up 3×3 tables. First we must make use of all R roll calls, not just those on which two men happen to both be present. For reasons of symmetry and logic, it seems reasonable to consider the cases in which A won and B was absent, or A was absent and B lost, as equivalent to "half victories" by A over B. Similarly cases in which A lost and B was absent, or A was absent and B won are equivalent to "half losses" for A over B. Thus we define a new power index of A over B, or P^* (A over B) as:

$$P^*(A \text{ over } B) = \frac{c + \frac{1}{2}b + \frac{1}{2}f - g - \frac{1}{2}d - \frac{1}{2}h}{R}$$

Note that P^* is transitive: if A is more powerful than B, and B is more powerful than C, then A will be more powerful than C according to P^*. Note also that $P^*(A \text{ over } B) = -P^*$ (B over A). Observe now that the numerator

$$
\begin{aligned}
&c + \tfrac{1}{2}b + \tfrac{1}{2}f - g - \tfrac{1}{2}d - \tfrac{1}{2}h \\
&= \tfrac{1}{2}(a + b + c) - \tfrac{1}{2}(g + h + i) \\
&\qquad\qquad - [\tfrac{1}{2}(a + d + g) - \tfrac{1}{2}(c + f + i)] \\
&= \tfrac{1}{2}(W_A - L_A) - \tfrac{1}{2}(W_B - L_B)
\end{aligned}
$$

where W_A is the number of roll calls on which A voted with the winning side, L_A is the number on which A voted with the losers, W_B is the number B won, L_B is the number B lost. But

$$\frac{W_A - L_A}{R} = \text{SS}(A) \text{ and } \frac{W_B - L_B}{R} = \text{SS}(B)$$

Hence by simple substitution we have:

$$P^*(A \text{ over } B) = \tfrac{1}{2}(\text{SS}(A) - \text{SS}(B))$$

Thus P^* is simply half the difference of success scores! Since the success scores can be computed easily for one man, the entire set of power indices between every pair of men in a legislature can be found easily.

In fact the legislators can be arranged in descending order of SS, such that if A is above B in the list, $P^*(A$ over $B) > 0$, or A is more powerful than B, and $P^*(A$ over $B)$ equals half the difference of SS. $P^*(A$ over $B)$ can range from $+1.0$ to -1.0. The P^* scores for each pair of legislators in Table 4.5 is zero.

Since legislatures are political decision making bodies (among other functions), the concept of success and its corollary, power, would appear to be the fundamental statistical approach to their analysis. American political science has tended to be so sociologically oriented that scholars have overlooked the simple political phenomenon of winning and losing.

Guttman Scaling

Historians ask why legislators cast their yeas and nays in the observed pattern. Correlation analysis that explains voting patterns on the basis of personal, party, or constituency characteristics (like the example in Chapter 3 on the Repeal of the Corn Laws) rarely is successful in explaining the entire pattern; the proportional reduction of error is rarely 100 percent. Naturally the historian turns to the hypothesis that each legislator has a more-or-less fixed attitude on each issue, and that he votes yea for measures that he approves, and nay on measures that are too strong for him to accept. The Guttman scale is a statistical device for discovering and describing situations in which all (or nearly all) the legislators vote according to a personal attitude on one funda-mental issue. If a roll call reflects the mixture of two different issues, and hence reflects the personal juggling and weighing that goes on in the mind of each legislator, simple Guttman scaling will not work. How-ever, techniques for multi-dimensional scaling are being developed that may eventually produce statistical procedures to handle this problem.

Consider a simple example. Suppose that a series of roll call votes are held on the appropriation of money for a particular government project. Assume that every legislator has in mind a particular maximum sum of money that he thinks should not be exceeded, ranging from zero (on the part of an enemy of the program) to $1,000,000 (on the part of its most fervent supporter). Note that these sums are *maximum* amounts for each legislator, not the *ideal* amount he would appropriate if he had the power. Suppose that the roll call votes are taken on the question of the exact amount to be appropriated. The first roll call is on the measure to spend $500,000, and it fails. The second roll call is to spend $300,000, and it passes. There are four possible responses each legislator could make. He could vote yea on both votes, thus indi-cating that his personal maximum was greater than $500,000. He could vote nay on both votes, thus indicating that his personal maximum was under $300,000. He could vote nay on the first, and yea on the second, indicating that his personal maximum was somewhere between $300,000 and $500,000. Finally, he could vote yea to the larger sum, and nay to the smaller sum. This last pattern is inconsistent with the hypothesis that everyone voted according to the single criteria of his personal maximum. If more than a few men showed the fourth pattern, Guttman scaling is impossible because some other factor must have been in operation. If very few men, or none, followed the fourth pattern, the historian can confidently divide the legislators into three groups— high, medium, and low. He has constructed a simple Guttman scale.

Figure 4.4   2 × 2 Table for Possible Voting Patterns of N Legislators on Two Roll Calls

		ROLL CALL #2		
		YEA	NAY	Σ
Roll call #1	yea	a	b	$a + b$
	nay	c	d	$c + d$
	Σ	$a - c$	$b + d$	N

When the Guttman scale is appropriate, it takes R roll calls and N legislators and divides the men into groups, with each group internally homogeneous, and each group more "pro" than the one *below*, and more "con" than the one *above*. Empirically, only a small proportion of all the roll calls generated by a legislature fit into a scale, and there may be several scales which emerge from one legislature's record. The problem of discovering which roll calls form a scale and which do not was a hit-and-miss proposition until MacRae set up a simple routine, complete with computer program, to handle the problem with a minimum of guesswork.

The MacRae method requires, first of all, the computation of the Q correlation coefficient between each pair of roll calls available. (Unanimous or near unanimous votes are usually discarded, but all the other votes, no matter what the motion was, may be used.) With R roll calls there will be $\frac{1}{2} R(R - 1)$ Q coefficients, each of which requires that each of the N legislators be classified into one of four categories. Obviously a computer or at least a countersorter is necessary if R exceeds 20, say, and N exceeds 50 or 75. To compute Q imagine the 2 × 2 table shown in Figure 4.4. Each of the N legislators fits into one of the cells—cell a if he voted yea on both roll calls, d if he voted nay on both, b if he voted yea on #1 and nay on #2, and c if he voted nay on #1 and yea on #2. (If some men were absent on either or both roll calls, tabulate only those who voted on both.) Then $Q = \dfrac{ad - bc}{ad + bc}$, and ranges from $+1.0$ to -1.0.

The advantage of using Q is that two roll calls with a high Q will scale because only a small proportion of the legislators will be in one of the cells—which will be the "inconsistent" cell corresponding to an inconsistent voting pattern. If three roll calls have a high Q with respect to each other, they will form a scale. Hence, if the historian sets up a matrix showing the Q values for each pair of roll calls, he will be able to find one or more scales simply by locating clusters of

**TABLE 4.6   Six Hypothetical Roll Calls for Ten Legis-
lators**

Y = yea; N = nay; O = absent

ROLL CALL	#1	2	3	4	5	6	(4-REVERSED)
Legislator							
A	N	Y	N	N	N	Y	Y
B	Y	N	N	Y	N	N	N
C	N	N	N	N	N	N	Y
D	N	N	N	Y	N	N	N
E	Y	O	Y	N	N	Y	Y
F	Y	Y	N	N	Y	Y	Y
G	N	Y	N	N	N	Y	Y
H	N	Y	Y	N	Y	Y	Y
I	N	Y	N	N	Y	N	Y
J	N	Y	N	N	N	N	Y
ΣY-N	3–7	6–3	2–8	2–8	3–7	5–5	8–2

roll calls each of which has a high Q with each other roll call in that cluster. Table 4.6 shows a hypothetical set of six roll calls, with ten legislators. Table 4.7 shows the matrix of Q coefficients calculated from Table 4.6. For example, comparing roll call #1 and #2, note that one man (Mr. F) voted yea on both (hence $a = 1$); two (Mr. C. and D) voted nay on both (hence $d = 2$); one (Mr. B) voted yea-nay (hence $b = 1$); five voted nay-yea (hence $c = 5$); and one (Mr. E) failed to vote on both roll calls, and so is excluded. Thus for #1 and #2,

$$Q = \frac{1 \times 2 - 1 \times 5}{1 \times 2 + 1 \times 5} = \frac{-3}{7} = -.43$$

Hence roll calls #1 and #2 have a Q of —.43. They do not scale. When dealing with a substantial number of roll calls, the matrix need not include those Q values ranging from +.50 to —.50, since only Q

values from $+.5$ to $+1.0$, or $-.5$ to -1.0 are of interest. If one requires that every intercorrelation inside a cluster exceed .6, and that the mean Q exceed .7, then good scales with less than 10 percent "inconsistent" votes will emerge. If one roll call has a negative Q coefficient from $-.6$ to -1.0 with every member of a cluster, the roll call should be "reversed"—that is, its yeas and nays interchanged. Reversal has the effect of changing the sign of Q, thus making the roll call eligible for inclusion in the cluster. Use only the reversed values of reversed roll calls when building the actual scale.

Inspection of the Q matrix in Table 4.7 shows that roll calls #2, #3, and #6 form a cluster. The Q coefficients between them are all $+1.0$. (Note that Q can equal $+1.0$ without the roll calls being identical; in fact $Q = +1.0$ when $b = 0$ or $c = 0$, and $Q = -1.0$ when $a = 0$ or $d = 0$.) Note that roll call #4 has a Q of -1.0 with #2, #3, and #6. Hence #4 must be reversed (see the last column in Table 4.6), and then it joins the cluster. Roll calls #1 and #5 do not meet the criterion of having a Q of at least .6 with every member of the cluster, nor do they form a cluster together. These two roll calls are outside the scale, and will not be considered further. Recall that they were used in the construction of the agreement matrix among legislators (see Table 4.2.)

TABLE 4.7   Matrix of Q Correlation Coefficients for
Six Hypothetical Roll Calls in Table 4.6

Since the matrix is symmetrical, only the upper half is given

	1	2	3	4	5	6	4 REVERSED
1	1	$-.43$	$.50$	$.50$	$.11$	$.46$	$-.50$
2		1	1	-1	1	1	1
3			1	-1	$.50$	1	1
4				1	-1	-1	-1
5					1	$.46$	1
6						1	1

The actual construction of the Guttman scale can now proceed. Each cluster of roll calls in the Q matrix forms a different Guttman scale. One extra step is useful. If two or more roll calls in a cluster are identical—that is, no one voted yea on one and nay on the other—then discard the surplus roll calls and save only one of the pair or trio.

The next step is to rewrite the scalable roll calls, ordering them not in chronological order (that is, 1, 2, 3, . . .) but in descending order of the number of yeas cast. (Or, if there are numerous absences, in order of percent yeas.) Thus the new order in the example is: #4 reversed (8 yeas, 2 nays), #2 (6–3), #6 (5–5), and #3 (2–8). The result is shown in Table 4.8.

TABLE 4.8   Rearranged Scalable Roll Calls

	4-REVERSED	2	6	3
Legislator				
A	Y	Y	Y	N
B	N	N	N	N
C	Y	N	N	N
D	N	N	N	N
E	Y	O = Y	Y	Y
F	Y	Y	Y	N
G	Y	Y	Y	N
H	Y	Y	N	Y
I	Y	Y	N	N
J	Y	Y	N	N

The four roll calls generate $4 + 1 = 5$ "scale types." Any voting pattern that is not identical with one of the five types indicates an "inconsistent" vote—which we now can call an "error." If absences are recorded, they should be changed to yeas or nays in such a way that the legislator conforms to one of the five types. This last step, of course, goes well beyond the data, and is risky and conjectural. It is not altogether necessary but it is done so that each man can be assigned a scale type. The scale type corresponds to the number of yea votes cast. There are numerous other possible types of voting responses, but only the five shown in Table 4.9 are consistent with the Guttman scale hypothesis. If Mr. C. had voted "yea" on roll call #3, his response (Y–N–N–Y) would not have fit any scale type, and so he would be inconsistent. The closest type to which he would fit, however, is still type 1, because he made only one "error" with respect to type 1, but more than one error with respect to each of the other types. The problem of assigning inconsistent legislators to scale types is delicate; indeed if the change of one yea or nay will not fit a man into one of the scale types, it is best to set him aside and call him "unscalable." The problem of absences is not too difficult, if no one recorded more than a few absences. We note that Mr. E was absent on roll call #2. If he was present and had voted nay, his response pattern (Y–N–Y–Y) would have been inconsistent. Therefore we assume that he would have been consistent, and assign him to type #4 on the reasonable basis that he would have voted yea on roll call #2. If a man shows a string of absences, he may be unscalable. Once the errors are "corrected" and the absences accounted for, the scale type of each man simply equals the number of "corrected" yea votes he cast. The array of patterns shown in Table 4.9 is called the Guttman scale. If a man can vote for #3, then he can approve and vote for 6, 2, and 4–reversed, and he belongs to type 4. If he draws the line between accepting 4–reversed, 2, and 6 on the one hand, and rejecting #3, he fits into type 3. And so on, so that the man who votes nay on 4–reversed, 2, 6, and 3—a

TABLE 4.9   Guttman Scale Types

ROLL CALL:	#4 REV	2	6	3	EXAMPLES
TYPE:					
0	N	N	N	N	Mr. B, D
1	Y	N	N	N	Mr. C
2	Y	Y	N	N	Mr. I, J
3	Y	Y	Y	N	Mr. A, F, G, H
4	Y	Y	Y	Y	Mr. E?

full fledged "anti"—fits into scale type 0. Further analysis then can be devoted to the type of men who constituted each type, and an analysis of the influences that led him there. Note the groups which emerge in a Guttman scale (for example, *B* and *D*, *I* and *J*, *A*, *F*, *G*, and *H*) resemble somewhat, but not exactly, the groups that emerged from the different technique of agreement analysis and taxonomic trees (see Figures 4.1 and 4.2). Thus, the historian looking at the Guttman scale would ask what Mr. A, F, G, and H had in common that put them between Mr. I and J on one side, and Mr. E on the other. Was it section, or party or person ality or constituency or what? And the historian is led to ponder, but now he has a very precise picture to study, and may ask very precise questions.

Measurement of Inequality and Segregation

Social structures characterized by the unequal distribution of some important quantity—wealth, income, power, education—often figure prominently in historical analysis. Several simple methods of measuring the total amount of inequality present in a given distribution are useful for summarization and comparison.

In elections and legislatures, where the 50 percent mark is critical to victory, the "minimal majority" or smallest grouping of units that will control 50 percent of the votes, is a convenient index. In legislatures or conventions where multiple committee membership is important, the smallest number of members who occupy 50 percent of all committee seats will at once measure the concentration of power and differentiate the "powerful" members from the rest. The malapportionment of legislatures, similarly, can be measured by finding the smallest population that potentially could control 50 percent of the seats.

In total populations the maldistribution of wealth, income, and education can be measured by finding what proportion of the total amount of a given resource is controlled by the richest 1 percent, 5 percent, or 10 percent of the population. For example in 1960 the top 5 percent of the families in the United States received 17 percent of the personal income going to all families, while the bottom 40 percent of the families also received 17 percent. In 1917, the top 5 percent received 25 percent of the total income, but, of course, the top families of 1917 were not identical to the top families of 1960.[1] This technique is valuable when data exist only for the most prominent (richest) members of a community. However, the focus on a small fraction of the population may obscure the true situation. Below the richest 1 percent equality might prevail (as was often the case among Russian serfs), or the amount of inequality might be just as high (as was the case among Southern slaveholders).[2] Two satisfactory measures of inequality which take into account the holdings of every unit—are the Lorenz curve and the Gini index.[3]

Consider the problem of measuring the inequality of average personal income among the sections of the United States. Table 4.10 shows the proportion of national population and national personal income accruing to each section in 1900. If perfect inter-sectional equality had existed, then the South Atlantic states, with 13.7 percent of the population, would have had 13.7 percent of the income; in fact these states received only 7.1 percent of the income, less than half of their "equal" share. Note that the data refer only to sections; they do *not* indicate the amount of inequality among individuals who lived in these sections.[4]

The first step in drawing the Lorenz curve and calculating the Gini index is to arrange the sections in ascending order, from poorest to richest on the basis of the ratio Y/P, as shown in Table 4.11. Two new values are shown for each section, $CumP_i$ and $CumY_i$, which are the cumulative subtotals of P and Y for all sections poorer than section i, together

TABLE 4.10   Sectional Distribution of Population and
Personal Income in the United States, 1900

P = population; Y = income; ratio = Y/P

				SECTION						
	USA	NE	MA	ENC	WNC	SA	ESC	WSC	Mtn	Pac
P	100%	7.3%	20.3	21.0	13.6	13.7	9.9	8.6	2.2	3.2
Y	100%	9.9%	29.0	22.4	13.3	7.1	4.9	5.2	3.1	5.2
Y/P	1.00	1.36	1.43	1.07	.97	.52	.50	.61	1.41	1.62

Source: Bureau of the Census, *Long Term Economic Growth: 1860–1965* (Washington, 1966), 70–71.

TABLE 4.11   Ordering of Sections by Relative Income
Levels

SECTION	Y_i/P_i	P_i	CumP$_i$	Y_i	CumY$_i$
1 = ESC	.50	.099	.099	.049	.049
2 = SA	.52	.137	.236	.071	.120
3 = WSC	.61	.086	.323	.052	.172
4 = WNC	.97	.136	.459	.133	.305
5 = ENC	1.07	.210	.669	.224	.529
6 = NE	1.35	.073	.742	.099	.628
7 = Mtn	1.41	.022	.764	.031	.659
8 = MA	1.43	.203	.967	.290	.949
9 = Pac	1.62	.032	.999	.052	1.001

with i. That is, CumP$_i$ for the West South Central section is $P_1 + P_2 + P_3 = .099 + .137 + .086 = .323$, and CumY$_3 = Y_1 + Y_2 + Y_3 = 0.049 + .071 + .052 = .172$. (The familiar Σ sigma notation can be used instead of Cum, but it tends to be confusing here.)

The Lorenz curve is drawn by plotting the nine points, CumP$_i$ and CumY$_i$ on a square graph, with CumP$_i$ (the population factor) always on the horizontal axis, and CumY$_i$ (the good that is distributed) always along the vertical axis, as shown in Figure 4.5. If equality had prevailed, then for each section $P_i = Y_i$, and hence CumP$_i$ = CumY$_i$. The line CumP = CumY is drawn in on the graph, and is called the line of equality. When perfectly equal distributions are plotted, they lie along this line. The more *unequal* a distribution is, the further its Lorenz curve lies *below* the line of equality. The area that lies between the Lorenz curve and the line of equality, therefore, is a visual representation of the amount of inequality present. Since the total area between the line of equality and the axes is $1/2 \times 1.00 \times 1.00 = .5$ (that is, the area of a triangle is one half the base times the height), the area of inequality can be anywhere from 0 to .5. The Gini index is simply twice the area of inequality, and thus can range from 0 (perfect equality) to 1.0 (total inequality).

The calculation of the Gini Index can be handled by first finding the area *under* the Lorenz curve, subtracting that from .5, and doubling the result. Figure 4.6 shows how the area under the Lorenz curve can be decomposed into N triangles and $N - 1$ rectangles (where N is the number of units, in this example nine). A little simple geometry shows that the area of the N triangles is equal to $1/2P_1Y_1 + 1/2P_2Y_2 + \ldots + 1/2P_nY_n$, and the area of the $N - 1$ rectangles is $P_2\mathrm{Cum}Y_1 + P_3\mathrm{Cum}Y_2 + \cdots + P_n\mathrm{Cum}Y_{n-1}$. A little algebraic manipulation therefore produces the formula for the Gini index,

$$ G = 1 - 2 \sum_{}^{n} P_i \, \mathrm{Cum} \, Y_i + \sum_{}^{n} P_iY_i $$

To find the Gini index it is necessary only to compute the values of P_iY_i and $P_i\mathrm{Cum}Y_i$ for each section i, sum these values and insert them into the formula. However, the values of Y_i/P_i also have to be known so that the units can be arranged in proper sequence from poorest to richest. Table 4.12 shows the computations—remember to keep the decimal point in the right place when mutliplying percentages!

TABLE 4.12 Calculation of Gini Index from Table 4.11

SECTION	P_i	Y_i	$\mathrm{Cum}Y_i$	P_iY_i	$P_i\mathrm{Cum}Y_i$
1 = ESC	.099	.049	.049	.0049	.0049
2 = SA	.137	.071	.120	.0098	.0164
3 = WSC	.086	.052	.172	.0043	.0148
4 = WNC	.136	.133	.305	.0181	.0416
5 = ENC	.210	.224	.529	.0470	.1112
6 = NE	.073	.099	.628	.0072	.0459
7 = Mtn	.022	.031	.659	.0007	.0145
8 = MA	.203	.290	.949	.0589	.1926
9 = Pac	.032	.052	1.001	.0017	.0320
Total	1.00	1.00		.1526	.4739

Hence $G = 1 - 2(.4739) + .1526 = .205$, qed.

Sociologists have developed a variety of measures of segregation, most of which are based on the Lorenz curve.[5] The relative difficulty of computing Gini indices, however, makes the "index of dissimilarity (or differentiation)" a more valuable tool for measuring how different two distributions are. The index of dissimilarity, D, can be computed rapidly and is usually highly correlated with the Gini index. Table 4.13 shows the distribution of Negro and white males by 10 occupational categories in Mississippi in 1960. The difference in percentages in each row is noted, the positive and negative differences are kept separate, and the sum taken of all the positive differences, which equals D. Note that D also equals the sum of the negative differences (rounding error makes a slight discrepancy.). Like G and other well behaved indices, D ranges from 0 to a maximum of 1.0.

Figure 4.5.   Lorenz Curve

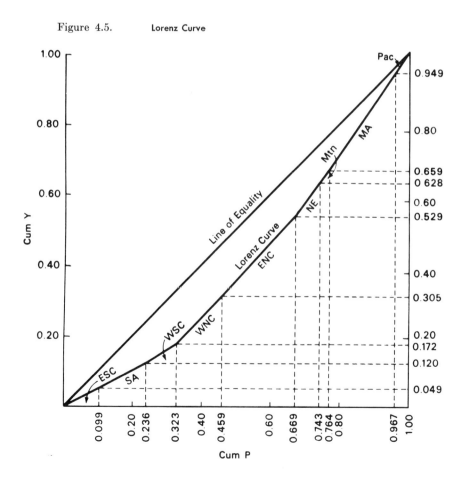

When used to measure residential segregation, D can be found without computing percentages. First, note which wards (tracts, precincts, or whatever) show a greater proportion of Negroes than the citywide proportion. Then add together the total number of Negroes living in just those blocks to get N, and then add together the total number of whites who also live on these same blocks to get W. Then $D = N/$total Negro population of city $-W/$total white population of the city.[6]

Time Series

A time series is a single variable whose values are recorded periodically. The most important elementary statistical methods for describing and

Figure 4.6.   Lorenz Curve

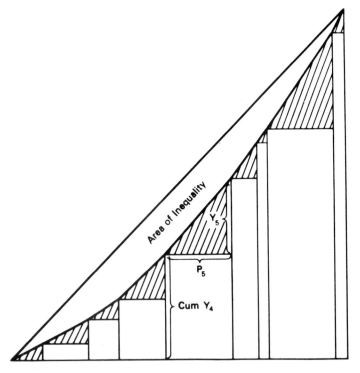

analyzing time series are *smoothing* out irregularities, checking for the existence of a *trend,* summarizing a trend, measuring *growth rates, correlating* two different series over the same years, and *interpolating* or estimating missing values. Of course, some preliminary checks are necessary before undertaking extensive analysis. The checks basically involve the question of comparability across time: Were the same definitions used every year? Did the territorial or membership base change? Were

TABLE 4.13 Occupational Distribution, Mississippi, 1960

CATEGORY	WHITE = X	NEGRO = Y	X – Y	X – Y
Professional, etc.	8.8%	2.3%	+6.5	
Farmers	13.2	18.4		−5.2
Managers	14.3	1.0	+13.3	
Clerical	5.9	0.9	+5.0	
Salesmen	7.6	0.5	+7.1	
Craftsmen	19.9	7.5	+12.4	
Operatives	18.7	18.1	+0.6	
Service workers	3.1	6.9		−3.8
Farm laborers	3.7	26.8		−23.1
Other laborers	4.8	17.5		−12.7
Total	100.0%	99.9%	+44.9	−44.8
	$D = $ .449 or .448			

Source: Jack Gibbs, "Occupational Differentiation of Negroes and Whites in the United States," *Social Forces,* 44 (1965), pp. 159–65, with data for 50 states, ranging from .218 in Maine to .624 in Wyoming. See Nathan Hare, "Recent Trends in the Occupational Mobility of Negroes, 1930–1960: An Intracohort Analysis," *Social Forces,* 44 (1965), pp. 166–73 for another example of the technique.

special seasonal factors involved that make careful attention to exact dates important?

The graphs of many time series, such as total population, are quite smooth, without sudden jumps and plunges, while other series have ragged graphs. Figure 4.7a shows the proportion of entries in *Writings in American History* (a comprehensive bibliography of scholarly and popular writing) classified by the editor as "biography" for the years 1913 to 1939. The analysis of shifting interests in historiography based on the data can be facilitated by smoothing out the oscillations so as to highlight the general pattern over time. Of course, particular fluctuations may be of some special interest—the jump in 1919 can be attributed to obituaries of Theodore Roosevelt, and that in 1932 to Washington's bicenteniary.

The simplest method of smoothing next to freehand smoothing is the three year moving average, which is simply the average value of the variable for a particular year, the preceding year, and the following year. For the data in Figure 4.7a the moving average for the year 1919 is one third the sum of the values for 1918, 1919, and 1920, or ($\frac{1}{3}$) (8.6% + 11.2% + 8.5%) = 9.4%. The value for 1920 is ($\frac{1}{3}$) (11.2% + 8.5% + 6.9%) = 8.5%, and the value for 1921 is ($\frac{1}{3}$) (8.5% + 6.9% + 8.7%) = 8.0%. Notice that it is not possible to compute a value for the moving average for the first and last years of a time series. The moving average for the biography percentage is shown in Figure 4.7b. Some fluctuations still remain, but the overall pattern is much more distinct.

To reduce fluctuations further, it is possible to take a three year moving average of the three year moving averages. The result is called a five year weighted moving average, since it equals the value for the central year multiplied by three, plus the values for the preceding and following years multiplied by two, plus the values for the next preceding and next following year,—all divided by nine to produce a value comparable to the original datum. Thus the five year weighted moving average for 1920 is ($\frac{1}{3}$) (9.4% + 8.5% + 8.0%) = 8.6%, which is equal to ($\frac{1}{9}$) (8.6% + 2 × 11.2% + 3 × 8.5% + 2 × 6.9% + 8.7%). The five year weighted moving average is also graphed in Figures 4.7a, 4.7b. Other possible weighting schemes lead to other kinds of moving averages, but the two described should be adequate for simplifying most time series and highlighting the overall patterns.

An important question in many research designs is whether a particular time series displays a tendency to grow (or diminish) over time, that is, whether it displays a trend. When the graph is smooth, simple inspection will provide the answer. When the graph is ragged, however, visual inspection may not be adequate, and the Spearman correlation test should be used. Simply rank the years from 1 (first year) to N (last year), and then rank the values from 1 (smallest) to N (largest). For each year, subtract the smaller rank from the larger rank, which produces N values of d. Square each of the d values, and add the squares together to get Σd^2. Spearman's rho (which is more thoroughly treated in Chapter 3) can be found by using the nomograph in Appendix C. Find the value of Σd^2 in the right-hand column, and the value of N in the center column. A straight line connecting these two points will intersect the left-hand column at the value of r_s, Spearman's rho. If r_s is high, a positive trend exists; if r_s is low, no trend exists. To see if r_s is low enough, use the following rule of thumb. For $N = 5$, r_s should exceed .5 before a trend is claimed; for $N = 12$, r_s should exceed .3; for $N = 30$, r_s should exceed .2; for $N = 45$, r_s should exceed .15; and for $N = 100$, r_s should exceed .1. (Note that the sign of r_s may be negative, indicating a downward trend; disregard the sign when using this rule of thumb.)

Occasionally it is important to ask whether a time series oscillates—that is, whether it is smooth or saw-toothed. Population series are smooth, but election turnout rates are typically saw-tooth, as shown by the turnout rates in New York state elections graphed in Figure 4.8a. An oscillating series suggests two possibilities. First, the series may be interpreted as the sum or composite of two different series, one consistently higher than the other. For example, turnout in presidential election years (1840, 1844, and so on) is almost always higher than turnout in off-years, and when all years are graphed, a saw-tooth effect is seen; each series by itself, however, is rather smooth (see Figure

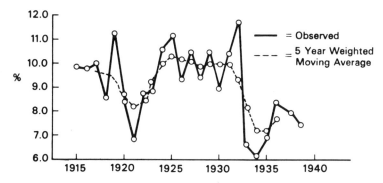

Figure 4.7a. Proportion of biographical entries in *Writings in American History*, 1915–1937

Figure 4.7b. 3-year and 5-year moving averages for data in Figure 4.7a

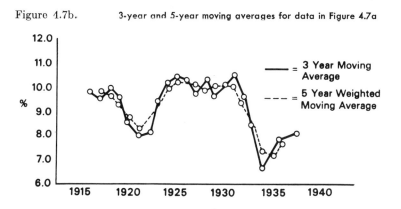

4.8b). The second possible interpretation hypothesizes a systematic negative feedback, or reaction. The share of the popular vote for one party,
for example, may be saw-tooth because voters turn against the party
in power in midterm elections. As a test for saw-toothness, simply count
the number of turning points in the graph. A turning point is a value
that is either higher or lower than *both* its neighbors. In a time series
composed of random numbers, with no trend and no tendency toward
oscillation or smoothness, two-thirds of the points (excluding the two
endpoints) are turning points, and one third are "in-between" points.
(This result may not at all seem obvious, but it can be proven rigorously using probability theory.) If more than one third of the points
are in-between, the series is smooth—in population series, there usually
are no turning points at all! If more than two thirds of the points
are turning points, then oscillation is present (In Figure 4.8a all the
points are turning points.). Strictly speaking, this test works only when
no trend is present; when a trend exists, it can be used by subtracting
the expected (trend) value from the observed values and graphing the
differences.

Time series with strong trends can be succinctly described in terms
of *growth rate,* either simple (arithmetic), or compound (geometrical,
exponential). A simple growth pattern shows a relatively constant increase of so many units per year. For example, the black percentage
of a city's population might increase at the rate of 1.2 points per year,
or 12 points per decade. In simple growth, the annual increase does
not tend to grow larger as time goes on.

A quick estimate of simple annual growth rate is $(B - A)/T$, where
A is the value in year 0, and B is the value in year T. When the
trend is not so strong, it is unwise to rely merely on the two endpoints
A and B; it is better to use all the values by means of a linear least
squares estimate, similar to the linear regression forms treated in Chapter
3. The fact that years are evenly spaced permits a simplification of
the estimates of the regression coefficients a and b in the form
$Y^* = a + bT$. The method for an *odd* number of years is set out in
Table 4.14. Note that year zero is taken as the middle year.

The procedure provides a trend estimate of the value of Y for each
year. The expected value Y^* for 1926 ($T = 0$) is $a + 17.01 \times 0 = a =$
138.8; the expected value for 1927 ($T = 1$) is $138.8 + 17.01 = 155.8$;
the expected value for 1921 ($T = -5$) is $138.8 - 5 \times 17.01 = 53.8$.
Notice that the expected values are very close to the observed values
except that in the boom year of 1929 there was more travel than expected, and in the depression year of 1931 there was less.

Simple growth rates (linear trends) appear as straight lines when
plotted on ordinary graph paper. Many time series, however, bend up-

ward when they are plotted—that is, the annual increase itself grows larger as time goes on. Population is a natural example—the net increase in population every year is not at all constant, but increases as the population grows larger—it is typically proportional to the size of the parent population. This kind of proportional (compound, geometric, exponential—the terms are similar) growth is measured by the average annual percentage increase, the compound growth rate r.

A quick method of estimating r is to estimate the length of time it takes the series to increase by 100 percent (that is, to double), by

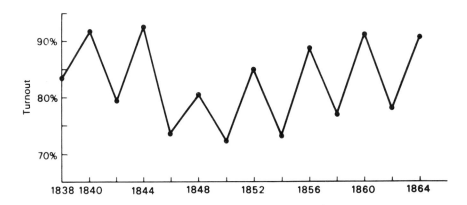

Figure 4.8a.   New York turnout, 1838–1864

Source: New York *Census of 1865* (Albany, 1865), lxxxii

Figure 4.8b.   Same data (unsmoothed)

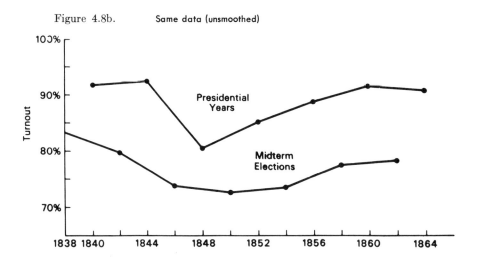

50 percent or by 25 percent. If a series doubles in T years, its average annual compound rate of growth $r \approx 70/T$. Thus if a population doubles in 25 years, it has a growth rate of approximately $70/25 = 2.8$ percent per year. If a series increases by 50 percent in T years its growth rate $r \approx 41/T$; if it increases by 25 percent in T years, $r \approx 22/T$.[7]

TABLE 4.14   Least Squares Trend Line; Y = billions of miles travelled by motor vehicles in the United States, 1921–31.

The regression coefficients are: $a = \Sigma Y_i/N$ and $b = 12\ \Sigma TY/(N^3 - N)$, where N is the total number of years; thus $a = 1527.1/11 = 138.83$ (the value for year $T = 0$), and $b = 12 \times 1870.9/1320 = 17.01$. (Appendix C gives the values of $(N^3 - N)/6$ that can be divided into $2\Sigma TY$.) The form is, therefore, $Y^* = 138.8 + 17.01T$, and the Y^* values are shown in the last column; notice how close they are to the actual values.

YEAR	T	Y = OBSERVED VALUE	TY	$Y^* = a + bT =$ EXPECTED (TREND) VALUE
1921	-5	55.0	-275.0	53.8
1922	-4	67.7	-270.8	70.8
1923	-3	85.0	-255.0	87.8
1924	-2	104.8	-209.6	104.8
1925	-1	122.3	-122.3	121.8
1926	0	140.7	0	138.8
1927	$+1$	158.5	$+158.5$	155.8
1928	$+2$	172.9	$+345.8$	172.9
1929	$+3$	197.7	$+593.1$	189.9
1930	$+4$	206.3	$+825.2$	206.9
1931	$+5$	216.2	$+1081.0$	223.9
		Sum $\Sigma Y = 1527.1$	$\Sigma TY = +1870.9$	

Source: U.S. Bureau of the Census, *Historical Statistics of the United States* (Washington: 1960), Series Q-322.

Series that gain a constant percentage every year appear as a straight line when their logarithms are plotted on simple graph paper. To save the trouble of looking up the logs, semi-log graph paper is available and should be used to plot any series that curves upward. Figure 4.9 shows the population of the United States and of Chicago, 1850–1940. The graph has four cycles, running from 1 to 10 (the bottom), 10 to 100, 100 to 1000, and 1000 to 10,000. Each cycle has been multiplied by 100,000, which means that populations between 100,000 and 1,000,000 can be plotted in the bottom cycle; populations of one million to ten

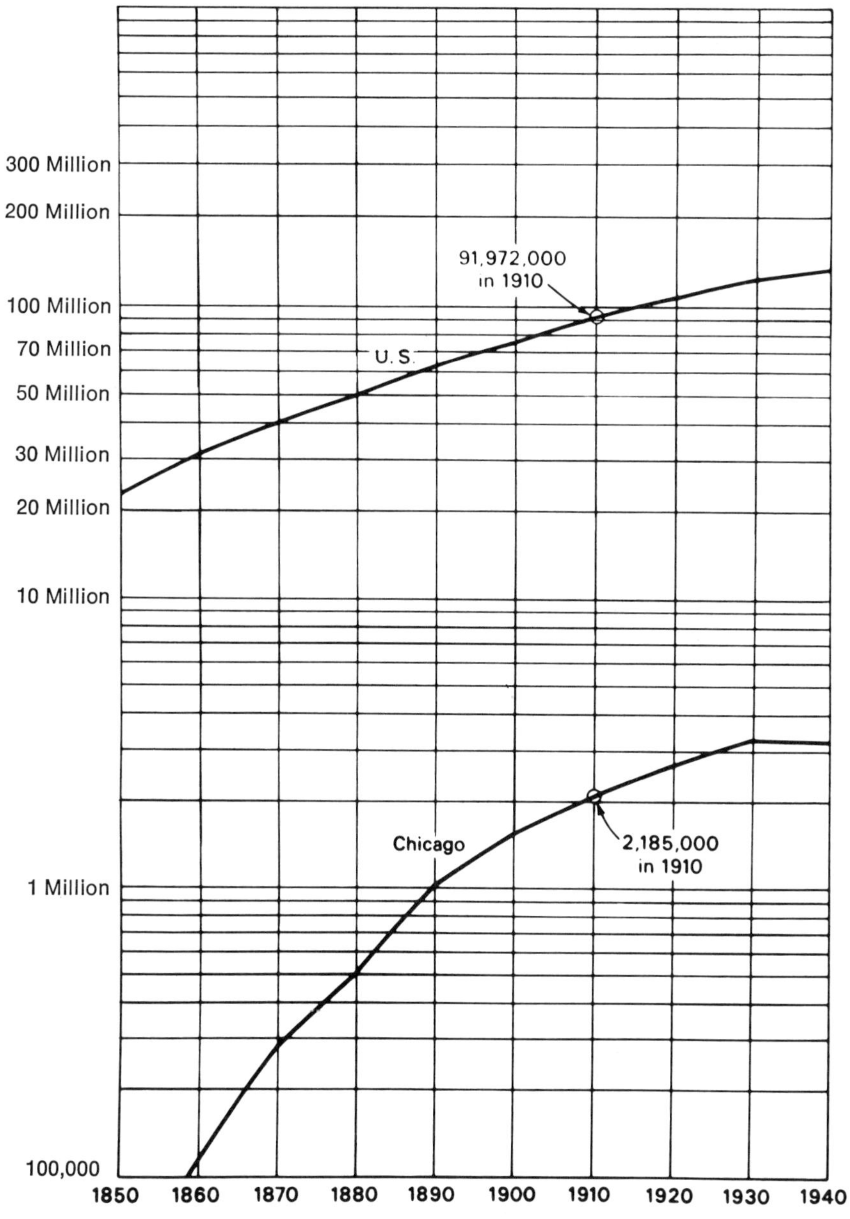

Figure 4.9.   Population growth of Chicago and the United States, 1850–1940

million in the next cycle; populations of ten to 100 million in the third cycle; and populations from 100 million to one billion in the top cycle. Note that in 1860 the population of Chicago was 109,000, while in 1940 the population of the entire United States was 132,000,000, and both totals are easily displayed on the same graph—something quite absurd on ordinary graph paper. Equal rates of percentage increase appear as parallel slopes.

Another growth pattern of some historical interest is saturated growth, which has an S-shaped graph (see Figure 4.10). Although the series grows exponentially for awhile, it eventually reaches its natural upper limit and levels off.[8]

Figure 4.10.

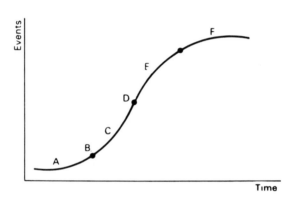

The correlation of two time series for the same years can pose very difficult problems. Visual comparison of the graphs (either on ordinary or semi-log graph paper) is adequate only when the series are highly correlated, in which case they will appear parallel. (If there is a high negative correlation, they will appear parallel if one is plotted upside down.) More precise technical methods are available, thanks to econometricians, but the correlation methods discussed in Chapter 3 cannot be used directly. The reason is simple: if two time series both have monotonic trends (that is, if the Spearman rho is not close to zero), then automatically they will be correlated, whether or not they are in any way related. Anyone who samples at random from a compendium of historical time series, will find most of the series are correlated with each other. (If the Spearman rho test for both series is close to zero, however, ordinary correlation, either Pearson or Spearman, can be tried.)

If either series shows a trend, it is wise to correlate not the observed values but rather the deviations from the trend. (That is, correlate $Y-Y^*$

with $X-X^*$ for the same years.) If the correlation then is positive, it means that both series tend to be above their trend, and below their trend, at the same time. Tetrachoric correlations are especially useful here. Let $a =$ the number of years both series are above their trend; $d =$ the number of years both series are below their trend; b and $c =$ the number of years one is above and the other below. The value of ad/bc or bc/ad, whichever is smaller, can then be entered in Table 3.5 to obtain the tetrachoric correlation between the two series. The values or deviations of series X may be correlated not with the corresponding values or deviations of series Y, but with the Y values one, two, or three years later—this is the phenomenon of lagged correlation, and may occur when the effect of one variable on another takes some time to become manifest. The exact lag may be of special interest, and the correlations can be computed with lags of one, two, and three years, first with X before Y and then, perhaps, with Y before X. Or if each series is graphed on separate sheets, it is possible to estimate the lag by moving the sheets until the graphs are parallel.

Many of the time series a historian uses, even the official ones that appear to be solid, contain, or did contain, gaps and breaks. Values for certain years often are missing, and have to be filled in by estimation or interpolation. Estimation from other sources is the indicated procedure whenever the gap overlaps a known discontinuity, such as war, depression, or domestic upheaval. If the gap covers a relatively quiet period, interpolation can be used. The simplest method is to plot the known values on a graph and visually estimate the trend, and sketch in the missing part of the curve. To estimate by linear interpolation the population of a state between (decennial) census years, the following formula will prove of use. Let the known (census) population in year zero be A, and the known population in year ten be B. Then:

the estimated population in year 1 = $.9A + .1B$
the estimated population in year 2 = $.8A + .2B$
. . . . .
the estimated population in year 9 = $.1A + .9B$.

If a long term trend exists and the series is smooth, the estimated value for year T should be derived from the trend equation. In more complex situations it is possible to interpolate one time series by using the known values of a parallel series.[9]

One of the basic problems every historian encounters is dating the occurrence of events. Precise dating and sequencing is necessary to sort out the causal structure of influences and events. If the era of Republican dominance began in 1896, the explanation for it is sought in the Mc-

Kinley–Bryan contest. If the era began in 1894, explanatory attention shifts to labor unrest; if the era began in 1893, attention shifts to depression. Of course, it is easy for historians to exaggerate the significance of the first occurrence of a development that takes many decades to unfold. Cities were less important in the mid-nineteenth century than the mid-twentieth century, but studies of urbanization tend to concentrate on the former period. It is also easier to use invalid indicators when dealing with time series. For example, families living in rented dwellings is distinctly an urban phenomenon, yet the proportion of American families in that category remained constant from 1890 (or earlier) to 1940 at the 54 percent level, then dropped sharply as a result of suburban trends. Yet the urban proportion of the population jumped from 35 percent to 70 percent at the same time. Clearly the homeowner/renter ratio, although useful in defining urbanization at any one time, is not helpful across time.

Quantitative methods may sometimes be of value, especially when dating social, economic, and political changes for which good data exist. Sometimes conditions change instantly (after assassinations and elections). More often change goes through clear stages (see Figure 4.10). Walt W. Rostow's stage theory of takeoff into economic development, although invalid as an explanation of economic history, remains very suggestive to the student of quantitative dating.[10] Stage A is pre-takeoff; nothing seems to be happening (although the perspicacious historian will undoubtedly discover the roots of all subsequent events here). At point *B* the first signs of upsurge appear, and during stage *C* the change is rapid. At point *D*, the "inflection point," mathematically speaking, the curve loses acceleration; it is still moving upward, but begins to slow down (stage *E*). Finally in stage *F* all of society that is going to be changed has been transformed and the curve levels off—or retrogresses—while awaiting a new type of social change with a new series of stages. Thus far only economists and a few sociologists have worried about the theory of dating changes using quantitative information; it is a topic that should appeal to the historian's imagination.[11]

FOOTNOTES

1.   U.S. Bureau of the Census, *Historical Statistics of the United States, Colonial Times to 1957* (Washington: 1960) series G132, p. 167, pro-

vides the basic time series; more recent data can be found in Bureau of the Census, *Statistical Abstract*.

2.   See I. D. Koval'chenko, *The Russian Serf Peasantry in the First Half of the Nineteenth Century* (Moscow: 1967, in Russian), which used a computer to find the distributions, but did not calculate indices of inequality; for these, see the review by Daniel Brower in *Journal of Social History*, 2 (1969), pp. 177–82. On slavery see Fabian Linden, "Economic Democracy in the Slave South," *Journal of Negro History*, 31 (1946), pp. 140–89.

3.   See Hayward Alker and Bruce Russett, "Indices for Comparing Inequality," in Richard Merritt and Stein Rokkan, eds., *Comparing Nations* (New Haven: 1966), pp. 349–72. A fascinating new index which can be decomposed to show the relative importance of regional, subregional, and individual inequality appears in Henri Theil, *Economics and Information Theory* (Chicago: 1967), pp. 91–134.

4.   For individual data see Ahmed Al-Samarri and Herman P. Miller, "State Differentials in Income Distribution," *American Economic Review*, 57 (1967), pp. 58–72; Simon Kuznets, "Distribution of Incomes by Size," *Economic Development and Cultural Change*, 11 (1963), pp. 1–80.

5.   For full details, and many historical examples, see Karl and Alma Taeuber, *Negroes in Cities* (Chicago: 1965), pp. 95–145; on the relationship of indices to the Lorenz curve see Otis Dudley Duncan and Beverly Duncan, "A Methodological Analysis of Segregation Indices," *American Sociological Review*, 20 (1955), pp. 21–27; for a thorough technical treatment see Otis Dudley Duncan, Ray Cuzzort, and Beverly Duncan, *Statistical Geography: Problems in Analyzing Areal Data* (Glencoe: 1961).

6.   See Taeuber and Taeuber, *Negroes in Cities,* pp. 236–38 for more details.

7.   More precisely, if the value in year 0 is A, and the value in year T is B, then $r = 2.3 \times (\log B - \log A)/T$. To save computational troubles, use the tables in U.S. Bureau of the Census, *Long Term Economic Growth: 1860–1965* (Washington: 1966), pp. 115–29.

8.   See Derek Price, *Science Since Babylon* (New Haven: 1961) and Daniel Bell, "The Measurement of Knowledge and Technology," in Eleanor B. Sheldon and Wilbert E. Moore, eds., *Indicators of Social Change* (New York: 1968), pp. 145–246, especially pp. 163–73, for fascinating applications to intellectual and social history. Technically the S-shaped curves are either Gompertz or logistic curves; for details see Frederick E. Croxton, Dudley J. Cowden, and Sidney Klein, *Applied General Statistics* (Englewood Cliffs: 1967), pp. 262–82; this textbook is excellent on elementary time series analysis.

9.   See Milton Friedman, "The Interpolation of Time Series by Related Series," *Journal of the American Statistical Association*, 57 (1962), pp. 729–57; and Myron Fiering, "On the Use of Correlation to Augment Data," *Journal of the American Statistical Association*, 57 (1962), pp. 20–32.

10.   W. W. Rostow, *The Stages of Economic Growth* (Cambridge, England: 1960); see W. W. Rostow, ed., *The Economics of Take-Off into Sustained Growth* (New York; 1963); for dating in general, see Arthur

Burns and Wesley Mitchell, *Measuring Business Cycles* (New York: 1946).

11.   Two valuable essays are W. E. Keeble, "Models of Economic Development," and D. Harvey, "Models of the Evolution of Spatial Patterns in Human Geography," in Richard Chorley and Peter Haggett, eds., *Models in Geography* (London: 1967), chs. 8 and 14. One technical note: a three year moving average may shift the apparent timing of a turning point either backward or forward.

FUNDAMENTALS
OF DATA
PROCESSING[1]

CHAPTER FIVE

Data Processing Defined

The term data processing is not part of the working vocabulary of many historians. Indeed, there are some historians who might argue that it probably has no legitimate place in that working vocabulary. Such an attitude stems from a misconception of precisely what data processing is. To some historians data processing suggests a machine—a mechanical monster—which robs history of its meaning and reduces it to the impersonal level. To others it implies an unmanageable mass of data that requires mechanical summarization.

Defined very simply, data processing is a systematic and orderly analysis of information in one form with the specific goal of converting it into another more usable form. Data may be construed as items of materials in any form whatsoever that relate to the goal of the inquiry. As discrete items, which may be of individual interest, data are independent of one another and singly do not permit a composite picture of interrelationships. Processing of such bits of information means their rearrangement according to a clearly defined procedure. Data processing, therefore, is purposeful handling of selected bits of information that produces knowledge not apparent in its original form.

The Data Processing Cycle

It must be emphasized that data processing fundamentally is a way of working with data. Neither the source (manuscripts, censuses, election returns, and so on) nor the quantity of the data (large or small) alters what may be called the data processing cycle. Every data processing application follows the same basic method of operation. This operation requires three steps. They are input, processing, and output. Input means the collected raw data arranged in a standard way in any form. Processing is the manipulation or rearrangement of the raw data according to some plan. Output is the presentation of new patterns distilled from the original data. Admittedly, this cycle may be oversimplified, but the point is that though historians may not describe this method of operation as data processing, this is precisely what we do. While the mechanics of application vary according to the various disciplines, every scholarly endeavor to produce new knowledge about empirical reality involves data processing.

Levels of Data Processing

The various levels of application of the data processing operation may be conveniently characterized as manual, mechanical, electromechanical, and electronic. It should be noted that at each level the human element of planning and controlling the operation is unchanged. On the other hand as the application moves from the manual to the electronic level, there is a corresponding decrease in reliance on human resources for the actual processing. A brief discussion of the four levels of data processing will clarify this point.

Manual data processing means that the individual physically handles the data and makes logical or mathematical decisions. The human brain makes these decisions in response to reading the data items or perceiving such empirical properties as size, shape, distance, or spatial configuration. The actual handling of the data is done with the paper and pencil.

Manual data processing can be illustrated with the following example. The manuscript registered voter list for precinct 17 of Tulsa, Oklahoma, for the election of 1922 contains the name, age, sex, occupation, race, and party preference of each registrant. If a composite picture of the registered voters of this precinct is desired, the first step is to copy the data for each registrant. Then these data are manually classified first for party preference and tallied. The same procedure is repeated for race, sex, occupation, and age groups. Next, percentages for each of the classifications are computed which present an overall view of the composition of the voters in that precinct. At each step of the manual data processing cycle—input, processing, output—the individual handles the data and makes logical and mathematical decisions. It is obvious that manual data processing can be extremely time-consuming and boring, and is highly susceptible to human error.

Mechanical data processing may be treated very briefly, since it differs from manual data processing in only one respect. In terms of the previous example, mechanical data processing uses a desk calculator, an adding machine, or slide rule to sum totals or calculate percentages. In both operations the same data are used, the same logical decisions are made, and the same results are obtained. However, in mechanical data processing a machine makes the arithmetical decisions. This reduces greatly the time expended and the likelihood of errors in calculations.

Electromechanical data processing differs sharply from the two methods previously described in that it requires data converted from the original document form into a machine readable language. The most widely used machine readable language is one in which the data are punched into the vertical columns of a card according to a specified code.[2] A machine then reads the data by sensing through electric impulses the presence or absence of punches. Thus in this operation a machine reads the converted data and makes such elementary logical decisions as classification and such mathematical decisions as counting. Computation of percentages still must be done by hand or with a calculator.

The previous example of precinct 17 in Tulsa, Oklahoma is useful at this point. The raw data consisting of the age, occupation, race, sex, and party preference of each registrant are punched in cards according to a uniform code, with one card for each registrant. These cards are then fed directly into a machine which rapidly sorts them, for example, into party preference and counts the number of each preference. The same sequence is repeated for sex, race, age, and occupation. In addition, subclassification such as determining party preference according to sex. race, age, or occupation is similarly obtained.

Why use electromechanical data processing? Machines are available which can process from 450 to 2000 cards per minute, a feature which greatly reduces the physical handling of the cards as well as the amount of time required to process the data. Also, counting is less prone to error and cross classification presents no problem. Even though electromechanical data processing is vastly superior to manual and mechanical operations, the only mathematical operation permitted is addition.[3] Furthermore, the processing cycle is not continuous, since each sequence interrupts the cycle when the data cards are fed into the machine.

Introduction of computers to data processing in the early 1950s marks the beginning of the fourth level of data processing, that of electronic data processing. This level of operation uses a computer, which, as Harold Borko puts it, ". . . *is* a machine that is able to calculate and perform sequences of arithmetical and logical decisions in accordance with preprogrammed instructions, thus eliminating the need for human intervention at each step."[4] The fascinating thing about computers is that they perform these operations at fantastic speed, usually measured in millionths and billionths of a second.

How does a computer work? First, the data are read into the computer by way of punched cards. The presence or absence of holes in the cards generates electric/magnetic impulses that transmit the data to a storage unit, where it remains until called to meet the needs of the processing operation. Through internal microminiature electronic circuits, transistors and the like, the data are manipulated in a sequence of steps according to instructions prepared by the human directing the operation. The results of the electronic data processing are then printed out.

A computer is a machine and no more, but like other machines it offers distinct advantages to its human user. Computer speed in responding to instructions prepared by the user is phenomenal. Also, computers, if so instructed, can make logical decisions of classification as well as perform calculations that range from simple multiplication and division to complex mathematical formulae. Furthermore, a computer with its built in checks offers greater accuracy in its operations than a human possibly can. It should be emphasized that the electronic data processing cycle is continuous. Once the data are ready and the machine is set up to operate, it takes over control with no interruption from the outside, save in the case of mechanical malfunction. This means that the data cards are handled only when placed in the machine and removed from the machine. Computer output, usually in the form of printed matter, contains the results of the computer's manipulations of the data according to programmed instructions. Printed output assures greater accuracy, since no other transcription is necessary, which in turn reduces the time required to complete the job.

The efficiency of electronic data processing can be illustrated with the previous example of data on registered voters in precinct 17 of Tulsa, Oklahoma. A computer, following essentially the same sequence of operations as those of electromechanical data processing plus multiplication and division, would process the data in less than five minutes. A sample page of the printed output is given below.

Figure 5.1.   Sample page of printed output of a computer analysis[a] of the registered voter list in precinct 17, Tulsa, Oklahoma

			VARIABLE NO.	1	AGE			
		VS.,	VARIABLE NO.	4	POLITICAL PREFERENCE			
TABLE SIZE =	8	BY	4					

	TOT	0	1	2	3	4	5	6	7
0	0	0	0	0	0	0	0	0	0
1	538	9	0	210	139	103	55	15	7
2	285	3	0	86	94	45	39	17	1
3	43	0	0	10	13	15	3	2	0
TOTAL	866	12	0	306	246	163	97	34	8

PERCENTS BY COLUMN FROM THE ABOVE MATRIX

1	529	0.0	68.6	56.5	63.2	56.7	44.1	87.5
2	282	0.0	28.1	38.2	27.6	40.2	50.0	12.5
3	43	0.0	3.3	5.3	9.2	3.1	5.9	0.0
TOTAL	854	0.0	35.8	28.8	19.1	11.4	4.0	0.9

PERCENTS BY ROW FROM THE ABOVE MATRIX

1	61.9	0.0	39.7	26.3	19.5	10.4	2.8	1.3
2	33.0	0.0	30.5	33.3	16.0	13.8	6.0	0.4
3	5.0	0.0	23.3	30.2	34.9	7.0	4.7	0.0
TOTAL	854	0	306	246	163	97	34	8

PERCENTS BY CELL FROM THE ABOVE MATRIX

1	61.9	0.0	24.6	16.3	12.1	6.4	1.8	0.8
2	33.0	0.0	10.1	11.0	5.3	4.6	2.0	0.1
3	5.0	0.0	1.2	1.5	1.8	0.4	0.2	0.0
TOTAL	854	0.0	35.8	28.8	19.1	11.4	4.0	0.9

[a] The analysis was made with the NUCROS program (modified for a 7040 IBM computer) described in Kenneth Janda, *Data Processing*, pp. 159-165. In the above table the headings for the columns (0-7) represent age groups in ten-year intervals. Entries in the 0 column mean that the age of some registrants was not given. The three rows represent political preference. The 1 denotes Democratic preference, the 2 denotes Republican preference, and 3 denotes Independent and Socialist.

Modes of Data Input

Thus far this chapter has briefly sketched the concept of data processing and its four levels of operations. The remainder of this chapter will focus on the input, process, output sequences for electromechanical and electronic data processing. The input sequence can be divided into the following sequential steps when electromechanical or electronic data processing equipment is employed: collection of raw data (this is discussed only tangentially since it is covered in detail in Chapter 1); coding of the data into a form suitable for machine "reading"; punching the data onto cards which are actually "read" by the machine; and controls to insure accuracy in the entire input process. The discussion of this sequence can be simplified if we orient it around the necessary requisites for machine analysis of data presented on punched cards.

Punched Cards

Almost everyone today is familiar with the punched card in one form or another. Punched cards are used widely in the form of payroll checks, government checks, utility bills, magazine subscription notices, and enrollment in colleges and universities, to name only a few. Although as early as 1728 a French engineer developed a loom that operated in response to perforated cards, the real father of punched cards was Herman Hollerith. Disturbed by the fact that tabulation of the 1880 census was not completed until 1887, Dr. Hollerith, a statistician with the Census Bureau, devised a new method for processing census data. The 1890 census data were recorded on cards by means of holes cut with a hand-operated punch. The cards were then processed through a tabulating machine that Hollerith designed. By using Hollerith's system, the 1890 census of 62 million people was completed in 1892 as compared with the 1880 census of 50 million people that required seven years to complete. Since 1890, use of the punched card gradually has expanded, so that very few people today are unfamiliar with it.

The rationale for punched cards is the need for efficient transfer of human language information into machine readable documents. Machine language consists of round or rectangular holes made in predetermined positions in specially designed cards. Punch cards usually are rectangular in shape with a fixed number of punchable columns and rows. A machine reads a punched card by means of the punches which complete electric circuits. Punched cards have advantages other than being machine readable. Once information is punched on cards and its accuracy verified, the cards form a permanent record and there is no need to use the original code sheets. Punched cards can be used repeatedly, even for purposes not initially considered. Also, they can be reproduced with no difficulty so that duplicates are always available. The speed with which punched cards can be processed, their uniform size, accuracy, and flexibility in use constitute a powerful research tool for the historian.

Two types of punch cards are available through the International Business Machines Corporation and the UNIVAC Division of Sperry Rand Corporation. Both cards are similar in size and shape, measuring 7–3/8 inches by 3–1/4 inches and .007 inches in thickness. The cards are made of high quality manila and can be used repeatedly. IBM cards are punched with rectangular holes, while holes in UNIVAC cards are round.

The IBM card is divided into 80 columns that are numbered from 1 to 80 from the left side of the card to the right. (See Figure 5.2.) Each column has 12 rows, which represent different punching positions, and may be used to record only one digit, 0 through 9, one letter of the alphabet, a through z, and special characters such as a plus sign (+) or a dollar sign ($). Descending downward from the top of the card, the rows represent +, −, and the numerals 0, 1, 2, 3, 4, 5, 6, 7, 8, 9. The +, −, and 0 are called "zone punches" and respectively occupy rows 12, 11, and 10. The top of the card is called the 12–edge and the bottom of the card the 9–edge. The positions in which cards are placed in a machine are designated 9–edge in or 12–edge in.

As noted previously, each card column may be used to represent a numeral, a letter of the alphabet, or certain special characters. Numeric punches require only one punch per column. For example, the number 9 can be recorded in any one of the 80 columns by punching a hole in the "9" row. Letters of the alphabet require double punches in a single column according to a special configuration. The letter a, for example, can be recorded in any one of the 80 columns through a combination of a punch in the "12" row and the "1" row. Most of the special characters require three punches per column. The important thing to remember is that each number, letter of the alphabet, or special character occupies a single card column.

The UNIVAC card is the same size and shape as the IBM card, but its column layout is different. The UNIVAC card contains 90 columns in which data may be recorded. This is accomplished through dividing the card into two sections of 45 columns each. The upper section is numbered 1 through 45 and the bottom section is numbered 46 through 90. Each section contains 6 rows designated 0, 1–2, 3–4, 5–6, 7–8, and 9. Numeric punches are divided into two types. Odd numbers and zero require only one punch per column. Even numbers require a combination of two punches, one in the row of the number in question and one in

Figure 5.2.   IBM punched card

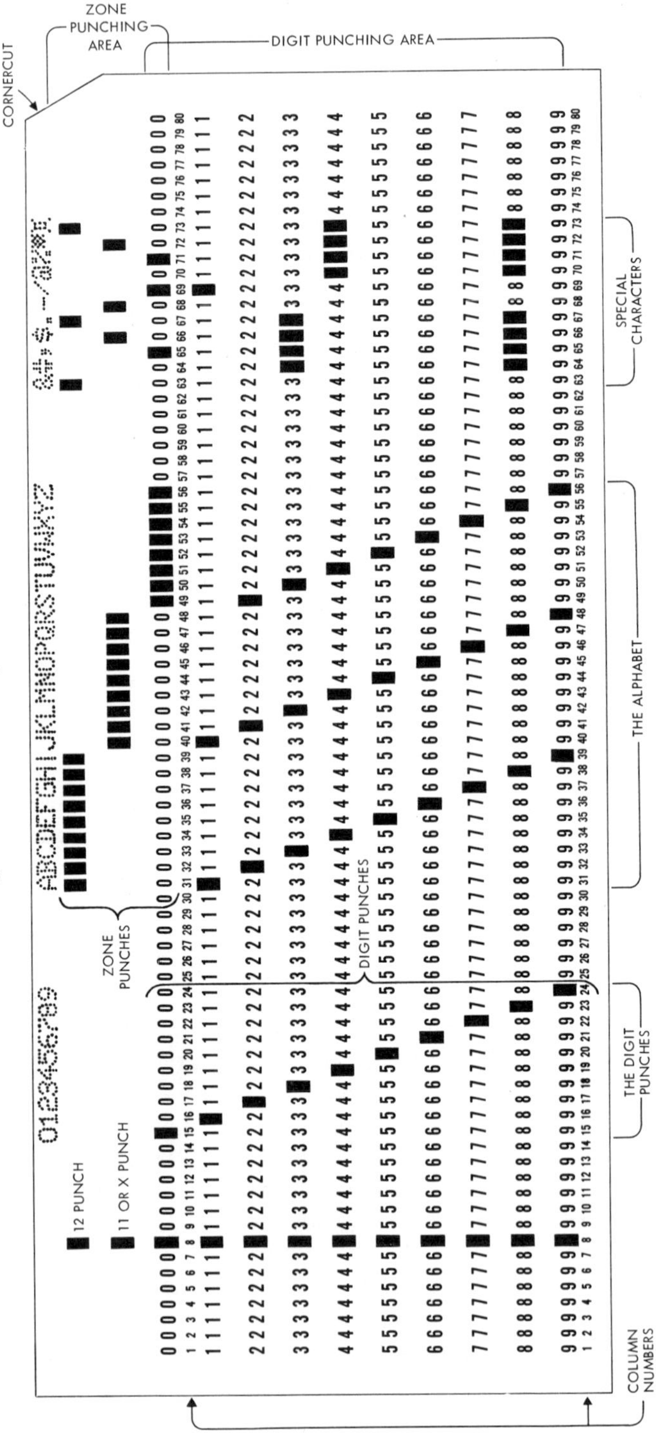

Courtesy of International Business Machines Corporation.

the "9" row. The alphabetic punch scheme requires a combination of three punches.

The decision of whether to use IBM cards or UNIVAC cards is dictated by the kind of equipment that is available to process the cards. Since IBM currently is the largest manufacturer of the data processing equipment, the 80 column card is used in most data processing operations. Accordingly, the remainder of this discussion of punched cards and machines will focus on the familiar IBM card.

Card Design

Since punch cards are a means of achieving efficient data processing, it is important that card design be given careful consideration in the early stages of planning of the project. Card design involves two key questions: (1) What information will be recorded on the card and for what purposes will it be used? (2) How may the required information be efficiently arranged on the card? The first question assumes that the researcher has developed a research design that requires clearly specified data. It is wasteful and inefficient to undertake card design without knowing what the goals are, what information is needed, and how it will be used. Even with a good research design, however, card design should begin with a careful examination of the data to be recorded. This examination might reveal gaps in the information which would limit its usefulness. On the other hand, previously unconsidered items, which offer another slant, might be discovered. In any event, the time invested in such an examination will be amply repaid in the most effective use of the punched cards.

It might be useful at this point to note that information may be categorized into identification data and computational data. The former identifies whether the information is about one or more persons, an event, or some other unit of inquiry. For example, the research design may require information about an individual, occupational groups, or a group of people in a geographic or political unit. Individual identification might consist of name, address, and geographical location, while occupational identification might include only geographic location and kind of work. Another kind of identification might refer to political units such as a precinct or a county. The importance of identification information is that it distinguishes both one card from another in a data set and one data set from another (for example, individual data and political unit data).

The second category of card information is computational data that will be used to answer the questions posed in the research design. Usually computational data are divided into categories most relevant to the goals of the investigation. These categories are usually called variables or attributes, and for example, might include age, occupation, religion, education, sex, race, and political preference for individuals. Another set of categories for voting behavior in presidential elections at the county or precinct level would include the number of votes each candidate received. In sum, the classification scheme employed is determined by the questions asked and the form of the raw data.

After deciding what information is required and the most appropriate classification scheme, the next step is that of determining the procedure which achieves efficient arrangement of the information of the card. Efficient here means maximum use of card space and uniformity in transferring information from the source document to cards. Uniformity in this transfer of information requires the construction of codes which determine the form in which information will be recorded. The setting up of codes is very important and is discussed later in this chapter.

One aspect of maximum use of card space is that of card "fields." A field represents the number of consecutive columns required to record certain information. Theoretically, the number of columns in a field may range from 1 to 80, depending upon the data to be recorded. If the data consist of numerals 0 through 9, then single column fields are used. However, if the data consist of two digits, for example 79, then two column fields are necessary. Suppose that county election returns are to be punched into cards and that the Republican return for a county is 987 votes. The digits 987 would be punched into a three column field and mean the number 987. The length of a field, or the number of columns in a given field, depends on the data and the codes used. The number of fields permitted on a single card is determined by the total number of card columns for all fields and cannot exceed 80.

A second aspect of maximum use of card space is card format. Essentially card format is the arrangement of identification data and computational data into the fields necessary to record the information so that similar information on other cards occupy the same columns. How does one go about laying out card fields? In considering the location of card fields, the following points should be carefully noted:

1. Initially the data should be examined to determine the maximum number of consecutive card columns that might be used for a given item of information.

2. Where possible arrange fields in the order that data are read from the source document. This reduces the likelihood of transcribing data into the wrong field.

3. As a general rule card space should not be wasted with blank columns. The exception is when fields occupy only a portion of card capacity and visual identification of fields is required.

4. Commas, dollar signs, and decimals usually are not punched. Instead, a practice called right justification is followed. This means that the last digit in numeric data is placed in the right-most column of that particular field. Each succeeding digit is placed in the next column to the left until the data are recorded. Any unused columns to the left are punched with zeros.

5. Similar information should be placed in identical columns in different sets of cards so that equivalent card columns on every card are punched according to the same format.

6. Provision should be made for adequate reference identifying information that distinguishes cards among themselves and permits merging into comparable data sets. If the number of fields exceeds card capacity and additional cards are needed, adequate identification should be entered on each card—for example, the data card for each person should be numbered.

A layout sheet similar to that in Figure 5.3 helps effective organization of the card format. In the format for voters in Tulsa, 18 fields have been laid out. Card columns 1 through 58 have been set aside for nine identification fields that range from 1 (item no. 1) to 20 columns (name and address) in length. Data fields begin with columns 59 and 60 which record age, while sex and race are entered in columns 61 and 62 respectively. Three columns, 63–65, are reserved for party preference. Home ownership and occupation identification occupy columns 66 and 67–68 respectively. Seven columns are set aside for value of taxable real estate (obtained from tax assessor's rolls) in the event a millionaire is included in the registrants. For most of the registrants only three to five of the seven columns will be used. An additional identification field in columns 76–79 permits sequential numbering of registrants up to a maximum of 9999 for each precinct. Column 80 makes possible a sub-sequence of cards for each registrant up to a maximum of nine.

What are codes and what do they do? Human language consists of symbols understandable by human beings. A code makes human language symbols compatible with a machine by converting the data from a human source document into machine language. It will be recalled that machine language is one or more holes in the 12 vertical positions in each column that a machine can sense. In effect, then, codes render human words and ideas into "machine-manipulatable and preservable records."[5] Codes can be expressed by alphabetic characters, numeric characters, or a combination of both. While alphabetic codes may be used, as a general rule they use considerable card space and require multiple punches in a single column, the latter of which poses technical disadvantages. The most effective code uses numbers to represent the original data. The use of state names as identification data illustrates this clearly. If an alphabetic code is used to identify state names, 14 columns would be used to record the longest state name. Even if abbreviations were used, five columns would still be necessary. However, if numerals are substituted for state names, only two columns would be used for the 50 states. This holds true for other qualitative data such as sex, marital status, political preference, occupation, yeas and nays, educational attainment, place of residence, and religion, to name only a few. Once a numeric code is constructed for data categories, the code will be entered into the same columns on each card. Thus the conversion of the original data into coded numeric form permits maximum use of punch cards as machine language records.

Figure 5.3   Coding Layout Sheet for Registered Voters in Tulsa, Oklahoma, in 1922

INITIAL COLUMN	FINAL COLUMN	NO. OF COLS.	ITEM	REMARKS
1	1	1	(1) Card type-individual demographic	ID
2	4	3	(2) Year	ID
5	6	2	(3) State	ID
7	9	3	(4) County	ID
10	12	3	(5) Precinct	ID
13	16	4	(6) Census tract area/enumeration district	ID
17	36	20	(7) Name	ID
37	58	22	(8) Address	ID
59	60	2	(9) Age	Data
61	61	1	(10) Sex	Data
62	62	1	(11) Race	Data
63	65	3	(12) Party preference	Data
66	66	1	(13) Home owner	Data
67	68	2	(14) Occupation	Data
69	75	7	(15) Taxable real estate value	Data
76	79	4	(16) Sequence number	ID

Construction and utilization of data codes call for careful thought, if maximum use of the original data is to be achieved. Traditionally, historical research is an individual effort with the researcher collecting and processing his data in terms of his own needs. Today there is an increasing awareness that the historian who collects and codes data should follow practices that permit wider use of the collected data by the academic community. Furthermore, the data a historian collects and codes might be of use to political scientists, psychologists, sociologists, and anthropologists for quite different purposes. Such an interdisciplinary approach and responsibility means, as the late Ralph Bisco has put it, that "As new concepts are developed, prior collections can be reexamined for evidence of change in the operation of individual or social processes."[6]

Bisco, former Technical Director for the Inter-University Consortium for Political Research, suggested several guidelines to follow in coding data that will contribute to effective interdisciplinary cooperation.[7] First, coding practices should be standardized within a data set. For example, if the data consist of legislative roll call votes, a given code for a yea vote or a nay vote must be followed consistently. A yea vote must be uniformly recorded, let us say, as a 1. Of course a 1 in one or more different fields could have an entirely different meaning. Also, coding practices should be as compatible as possible with codes used in other data collections. Standardized coding conventions assure this compatibility and mean that other researchers can use the coded data with little or no reformating. Currently the Inter-University Consortium for Political Research is developing such a standardized code for historical data.[8]

A second guideline is to retain as much of the original data as possible in coded form. Suppose that age of individuals is an item of information to be recorded on cards. One method of coding the data is to set aside one column in which age groups are coded as follows: one to nine years—0; 10 to 19 years—1; 20 to 29 years—2; 30 to 39 years—3; 40 to 49 years—4; 50 to 59 years—5; 60 to 69 years—6; 70 to 79 years—7; 80 to 89 years—8; 90 to 99 years—9. A far better code that preserves the original age of each person uses two columns in which to record the actual age. For example, a person 31 years old would be coded as 31. Another example is that of handling income or value of real estate. Instead of using a one column field with a code of 0 to 9 for ten different income groups, the actual dollar value is recorded in a seven column field, which will accommodate sums up through $9,999,999. Recording as much of the original data as possible reduces coding error and increases coding speed, since the coder does not have to evaluate the data. Equally as important is that with the original data preserved the coded information is of greater use to other researchers.

Another guideline is that when the range of data permits it, two column fields should be used and a code devised that makes the first digit analytically useful. Level of educational attainment, for example, requires a two column code field. The first digit indicates a general attainment level—elementary, junior high, senior high, college, and so on—while the second digit specifies a particular year or grade within a level. A 0 in the first column indicates a person did not go beyond the elementary level and 1 through 6 in the second column specifies grades one through six. A sixth grade education is coded 06. Since this code follows a natural sequence of order, only the highest level of attainment is recorded.

A fourth guideline for coding data is that of permitting only one punch per column. This restricts the code to 0, 1, 2, 3, 4, 5, 6, 7, 8, 9. If more than ten categories are needed, then a two or three column field is used. Multiple-punches, or alphabetic codes, severely limit the usefulness of the coded data. If the cards are processed in a counter sorter, the operation takes twice as long and becomes rather involved. More important, however, is that virtually all computer analysis is geared to single or numeric punches. The only advantage of multiple-punched data is that of recording more data per card. Since IBM cards cost very little, the disadvantages of multiple-punches nullify the space-saving advantage.

In some instances alphabetic punches are necessary in identification data such as a person's name and address when a Social Security number or some other identification number is unavailable. It will be recalled that in Figure 5.3 two fields of 20 columns each were reserved for name and address. This was done because the data for individuals were recorded by name and address in the original documents. In this instance the use of names and addresses permitted visual cross checking of the two sources and insured recording of the correct information for each person.

A fifth guideline that Bisco suggests is ". . . never to use blank (or space) as a code category."[9] Blank spaces are easily confused with unused columns and a coder might unwittingly enter data in a column because it appears not to have been used. Also many statistical routines for computer analysis equate blanks with zero punches. Of course if names and addresses are recorded in alphabetic characters, then a blank serves to separate words.

It is of some consequence that a code book be kept which specifies card fields and clearly explains what each code represents. While a machine or computer senses numeric punches, a human being must assign meaning to them. A code book, therefore, preserves the unique meaning of each code. To rely on memory is hazardous, especially after some time has elapsed since last working with the data. Furthermore, a code book is indispensable if the coded data are made available to other researchers.

The code description in Figure 5.4 for individual demographic data at the precinct level for Tulsa, Oklahoma is based in part on the coding conventions developed by the Inter-University Consortium for Political Research and the authors' own innovations. An asterisk indicates the ICPR code.

Several suggestions about the mechanics of data coding are in order. If the information in certain fields is repeated on each card, for example year, state, and county, record this on the first line and write duplicate under each heading. This notifies the keypunch operator, who works from the data code sheets, that the machine can be set to duplicate these columns automatically. Duplication like this, of course, reduces the amount of time required to punch the cards. It is advisable that a soft lead pencil be used on the code sheets, since errors can be easily erased and the correct data entered. This eliminates uncertainty about which is the correct entry, the scratched out code or the miniscule code in the unused space. Another good practice that reduces uncertainty

Figure 5.4   **Code Description for Individual Demographic Data**

COLUMNS	INFORMATION		CODE
1	Data for individuals		1
2–4	Year (only the last three digits)*		920
5–6	State code for Oklahoma*		53
7–9	County code (three-digit codes developed by the Bureau of Census are standard)		072
10–12	Precinct number (obtained from official election returns)		001 to 999
13–16	Census tract or enumeration district in which the individual lives (obtained from Census Bureau maps)		001 to 999
17–36	Name (last name first and initials)		
37–58	Address (street and number)		
59–60	Age*	Not given	00
61	Sex*	Male	1
		Female	2
62	Race*	White	1
		Negro	2
63–65	Party preference*	Not given	000
		Democratic	100
		Republican	200
		Independent	328
		Prohibition	361
		Progressive	370
		Socialist	380
66	Home owner (obtained from tax rolls)	Yes	1
		No	2
		Not given	0
67–68	Occupation†		
69–75	Assessed value of taxable property		Actual value in dollars 0001 to 9999
76–79	Card number		
80	Blank		

* ICPR Code

† The numeric code is to lengthy to list here. It is the two-digit occupational classification of the Department of Labor.

is to code the letters i and o as I and ∅, so as to distinguish them clearly from 1 and 0.

The preceding example of code construction is only one aspect of codes for historical data. The reader desiring a variety of examples may examine with profit Kenneth Janda's handbook, *Data Processing.*

Information from code sheets normally is transferred to punch cards by means of a keypunch machine which requires an operator to read the code sheets and record the data by way of a keyboard similar to a typewriter. The most widely used keypunch machines are the IBM 26 Card Punch and the IBM 29 Card Punch. The main difference between the two is the number of characters which may be punched. The IBM 26 48 character set includes the alphabetic and numeric characters and 12 special characters. The IBM 29 punches the alphabetic and numeric characters plus 24 special characters. The latter character set is designed for use with IBM 360 computers. The IBM 26 is pictured in Figure 5.5.

Operation of a keypunch is quite simple. Anyone who types, even with the "hunt and peck" system, will quickly master its mechanics. Figure 5.6 displays the keyboard which contains keys for numbers, letter, special characters, and specific machine operations. Unlike a typewriter, the shift key does not punch upper and lower caps. Instead, each key contains two different characters, and when the numeric shift key is held down, the top character on each key is punched. For example, pressing the key 3-O punches 3 when the numeric key is depressed, but an O when the alphabetic shift key is down.

Cards pass through the machine as illustrated in Figure 5.7. They are placed in the card hopper face down with the 9 edge in and move down from the hopper when the feed key is pressed. When the register key is pushed, the card moves into position so that the first column is under the punching station ready to be punched. Pressing key U-1 with the numeric shift down causes a 1 to be punched in column 1 and automatically the card spaces to column 2. This procedure is followed for each column. If a column is to be left blank, the space bar is used. When column 80 has passed the punching station, the card is released and moved to the reading station. At the same time another card is fed from the hopper. The register key is punched again and the card moves to the punching station. The card at the reading station and the punching station move simultaneously column for column. If there is information on the first card that is to be repeated on the second card, the duplication key is pressed for the appropriate columns. As soon as column 80 on the second card passes the punching station, a third card moves down from the hopper and the first card moves to the card stacker.

If only one card is to be punched, it is placed by hand in the punch station position and the register key is depressed. The data are then punched in, after which the release key is pushed. This key advances the card to the next station without having to move a single column at a time through column 80. Depressing the key moves the card to the card stacker.

A single card may also be duplicated. Suppose that after all the cards are punched an error is discovered. A 1 was punched in column 61 instead of a 2. It is not necessary to repunch the card a column at a time. Insert the card with the error in the reading station section and a blank card in the punching station section. Press the register

Figure 5.5.   **IBM 026 Keypunch**

Courtesy of International Business Machines Corporation.

Figure 5.6. Keyboard of the IBM 026 Keypunch[a]

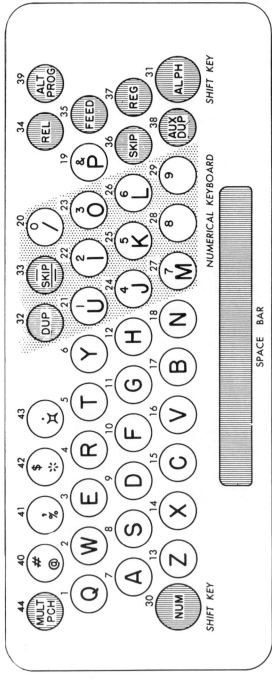

Courtesy of International Business Machines Corporation.

[a] This keypunch with a 46 character set can be used with any IBM computer except the 360 series. It requires use of the 029 keypunch with a 256 character set.

key and then hold the duplicate key down until column 61 on both cards is reached. A column counter located just above the reading station registers the number of each column as it passes through the two stations. The counter consists of an IBM card wrapped around a spindle whose movement is geared to that of the two cards. When column 61 reaches the punching station, the spindle will have revolved so that the column is under a vertical pointer. At this point release the duplicate key, press the numeric shift key and punch in a 2, the correct code. Then press the duplicate key until column 80 moves through the punching station. Pushing the release key moves the corrected card across the reading station and into the card hopper.

Figure 5.7. **Path of a card through the punch**

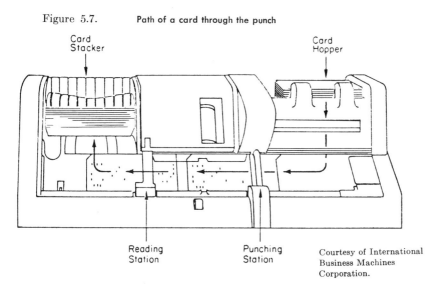

Card
Stacker

Card
Hopper

Reading
Station

Punching
Station

Courtesy of International
Business Machines
Corporation.

Is it necessary for the historian to learn how to operate a keypunch machine? In terms of day-to-day keypunching of voluminous data the answer is no. Every computer center has keypunch operators who are as fast and accurate as highly skilled typists. Most keypunching should be left to these operators. However, there are occasions when an error is discovered in a card in the middle of a job. To turn the mispunched card into the keypunch section for correction would mean an unnecessary delay. On the other hand, if the user knows the basic operations of keypunching, he can quickly make the correction. Thus, even a rudimentary knowledge of how to operate a keypunch is highly desirable.

Some historians may question how accurate punched cards are. This certainly is a problem but it is not unique. Misunderstanding of the context of a historical document, misinterpretation of the document itself, assignment of authorship to the wrong person, and reliance upon other faulty data, to name only a few, may be included in a list of errors historians have committed. As in other endeavors, the problem is that of keeping errors at a minimum level. One way to spot check for accuracy of keypunching is to compare the characters printed at the top of every tenth card with the information on every tenth line on the coding sheets. The most accurate verification, however, comes from repunching the cards on a IBM 56 verifier.

This machine resembles the IBM card punch in appearance and operations. The punched cards are placed in the card hopper and moved to the verifying station. The operator then repunches the information from the coding sheets the card punch operator used. In the verification process, the cards are not actually repunched. Instead, the verifier senses the punches in the cards and compares them with those registered by the operator. For example, if a 2 is coded for column 16 and the verifier hits that key, the hole in column 16 is compared. If a 2 was originally recorded, then it is assumed that the entry for that column has been correctly punched. If some other character was originally punched in column 16, then the verifier automatically places an "error notch" for easy correction on the 12 edge of the card directly above that column. If all the verification punches agree with those in the card, a notch is placed on the right-hand side of the card.

The accuracy of the verifier assures a user of correctly punched cards according to the coding sheets. Most computer centers have a verifier in the keypunch section and its use should be requested if more than 25 cards with 80 columns of information are to be punched.

Reproducer

As noted earlier in this chapter, one of the chief advantages of using punched cards is their reproducibility. The keypunch can be used to duplicate cards column by column, but this is very time-consuming if the number of cards to be duplicated is large. IBM manufactures a machine called a reproducer, which transfers up to 80 columns simultaneously from one card to another. The IBM 514 Reproducing Punch, pictured in Figure 5.8 reproduces cards at the rate of 100 per minute. The basic principle of operation is similar to that of the tabulator. The electrical impulses generated by punches travel to a control panel which directs the machine to punch out the appropriate row in each column. The control panel can be wired to reproduce up to 80 columns of information. Again it is unnecessary to learn control panel wiring techniques, since most computer centers have prewired panels for 80 by 80 reproduction. The 80 by 80 designation means that all 80 columns on the original cards will be transferred into the same 80 columns on blank cards.

Accuracy of reproduction is assured by comparison of each original card with each punched card. If a column on the original card has been reproduced incorrectly, an electrical circuit will be completed and the machine will stop. A comparing unit indicator will light up that shows which column or columns have been incorrectly reproduced. Even if no incorrect punches are indicated, it is a good practice to compare visually the first and the last cards of the original cards with the first and last cards of the reproduced cards. This is done by placing the first card of the original deck on the first card of the duplicated cards, aligning them carefully, and holding them up in front of a light. If the original card has been correctly reproduced, you should be able to

Figure 5.8.   **The IBM 514 Reproducing Punch**

Courtesy of International
Business Machines
Corporation.

see light through every punched column of the original card. The same procedure is repeated for the last card.

The Interpreter

While reproduction of a set of cards produces cards punched identically as the original set, the printing is not duplicated. This makes interpretation of the contents of the unprinted cards a difficult task.

Figure 5.9.  IBM 557 Interpreter

Courtesy of International Business Machines Corporation.

IBM has developed interpreting machines which translate the holes in a card into printed information on the same card. The most widely used interpreter is the IBM 557 (which interprets 100 cards per minute) shown in Figure 5.9.

The interpreter can print a maximum of 60 characters at a time. To print 80 columns of information the cards must be fed through the interpreter two times. The first run prints the information at the top of the cards, and the second run prints the remainder of the information between the 11 and 12 punch positions. A dial on the machine controls the line on which the information is printed. A control panel determines which columns will be printed. A prewired panel that prints 60 characters on the first pass and 20 characters on the second pass probably is available at most computer centers.

Operation of the interpreter is quite simple. Check the control panel to make sure the proper panel is in place. Set the printing position knob at the upper line. Place the cards into the hopper face up with the 12 edge in. Then press the start button. The machine will automatically interpret the cards at the rate of 100 per minute until all the cards have been processed. The same procedure is repeated for the second pass except that the print position is set at the lower position.

This treatment of card-processing machines has been designed to acquaint the reader with the basic machines available at most computer centers so that when he visits the center there will be an initial familiarity. Learning how to operate each of the machines is largely a matter of actually using them. A careful review of the discussion in this chapter along with actual practice in operating the machines will soon dispel any mysteries about card processing machines.

Card Handling and Storage

Since punched cards constitute a permanent source of data, they should be stored and handled properly. Punched cards may be stored in boxes, but if the boxes are stacked, the pressure may bend or crush the edges of cards in the lower boxes. For this reason it is far better to store cards in metal trays. Also, rubber bands and paper clips to bind small decks of cards should be used sparingly, since over a period of time, they tend to cut the card edges. Warped or bent cards will not feed properly into the machines and frequently cause the machines to jam. This can result in lost cards and lost time. It is always a good practice to check a deck of cards for rough edges before placing them in a feeder hopper. If a frayed edge is detected, the card can be quickly duplicated on the keypunch. Also it is a good practice to joggle the deck of cards lightly on a special plate attached to each machine to insure that the cards are properly aligned. Then fan the cards quickly to remove any static electricity that might cause cards to stick together.

Modes of Processing and Output

Up to this point attention has centered on converting human language information into machine language on punched cards. Now we turn to machines that process punched cards.

The Counter-Sorter

The most common data processing machine is the sorter. It interprets the holes punched in a single card column and sorts into groups cards having the same punch. A sorter with a counter will sum both the number of cards in each group and the total number of cards processed.

Several companies market counter-sorters, but probably the most widely used ones are manufactured by IBM. Either the IBM 82, which sorts and counts 650 cards per minute, or the IBM 83, which sorts and counts on a single column at the rate of 1000 cards per minute, will be found at most computer centers. The IBM 83 is shown in Figure 5.10.

Figure 5.10.   **IBM 83 Counter-Sorter**

Courtesy of International Business Machines Corporation.

As a card passes between a sensing device or wire brush and a metal roller, the card as an insulator prevents the completion of an electrical circuit. The circuit can be completed only when a hole permits contact between the wire brush and the roller. If there is no hole in a column, no circuit is completed and there is no electrical impulse. The movement of the card between the two contact points is designed so that a different impulse is created for each of the 12 punching positions. Accordingly, the machine easily distinguishes between, say, a 1 punch and a 9 punch, and a different operation occurs with each punch.

The wire brush is movable and can be positioned at any one of the 80 columns on a card, one column at a time. Once a card is read, it can be sent to any one of the 13 pockets. These correspond to the 12 vertical punching positions (12, 11, 0, 1, 2, 3, 4, 5, 6, 7, 8, 9), plus a reject pocket which receives cards that contain no punch in the column being sorted on. A bank of switches permits suppression of sort on any of the 12 positions. For example, if you wish to sort out only those cards containing a 1 in column 16, all the switches except the 1 switch would be pressed. All the cards not containing a 1 in that column would be sent to the reject pocket, while those containing a 1 would be sent to the 1 pocket. This is particularly useful in retaining the original order of the other cards, which will stack up in the reject pocket as they were placed in the card hopper.

In addition to the wire brush, and the digit suppression keys, there is a sort-selection which permits five sorting patterns: numerical, zone, alphabetic-sort 1, alphabetic-sort 2, and alpha-numerical. With the numerical switch (N) on, only numerical punches or single punches are sorted into pockets 0 through 9. All other punches go into the reject pocket. The zone switch (Z) sorts 0, 11, 12, or alphabetic punches into their respective pockets, while cards without a zone punch fall into the reject pocket. Alphabetic-sort 1 (A-1) sends cards with the letters A through I to pockets 1 through 9, while cards with letters J through R and S through Z go into the 11 and zero pockets respectively. Alphabetic-sort 2 (A-2) switch sorts either letters J through R or S through Z into their proper pockets. All other cards go into the reject pocket. The Alpha-Numerical (A-N) position sorts digit punches into pockets 0 through 9 and letters A through I and J through R into pockets 12 and 11 respectively. Letters S through Z go into the reject pocket.

A step-by-step operation of the counter-sorter can be illustrated with the punched cards for registered voters in precinct 17. Suppose we want to sort the cards according to political preference. The cards are placed in the hopper face down, 9-edge in. The sort-selector switch is set at N and the reading brush is moved to column 63, which contains the first digit of the political preference data. The counter is checked to be certain it is set at zero. The start button is pushed and the cards move one at a time between the roller and the wire brush until all the cards are sorted and counted. The number of cards falling into the "1" and "2" pockets represent the number of registered Democratic and Republican voters. Since the cards falling into the "3" pocket may include several party preferences (Independent, Prohibition, Progressive, and Socialist), a second sort is required for these cards. In this sort the reading brush is set at column 64, the second digit of the political preference code. In precinct 17 there were 866 registered voters, of whom there were 538 Democrats, 285 Republicans, 24 Socialists, and 19 Independents. If a further breakdown of each group is desired, each stack of cards is taken from a pocket and placed on separate stacks on top of the machine. Sorts and counts could then be executed on age, race, and occupation for each political preference group by setting the reading brush at the proper column.

After completion of these sorts, the original sequential order of the cards is lost. This is easily restored by sorting the cards on each column in the sequential identification field, beginning at column 79 and working leftward to column 76. After each sort, the cards are removed from the 0 pocket first and followed in order by those from the 1, 2, 3, 4, 5, 6, 7, 8, 9 pockets. They are placed in the hopper with the 0 cards on the bottom followed by cards from pockets in ascending order. After four sorts the cards will be sequentially arranged, beginning with card number 1 and ending with card number 868.

Sorting on alphabetic punches is more complicated and time-consuming than sorting on numeric punches. Application of the operation outlined above to alphabetic punches requires at least two sorts on each column and a very careful handling of the cards. It is for these reasons that codes consisting of alphabetic characters are strongly discouraged.

The Tabulator

The next major piece of equipment usually found in a computer center is a tabulator or accounting machine. Though the tabulator can add, subtract, and accumulate, its chief function in most computer centers is to print the contents of cards, an operation called "listing." While IBM markets tabulators with varying capacities, model 407 probably will be found at most computer installations. Model 407, which prints up to 120 lines per minute (one line per card), is shown in Figure 5.11.

The tabulator operates automatically through a control panel which directs the machine to perform various functions. For example, a simple listing of cards requires a special control panel, as do addition or subtraction. The wiring of control panels is quite complicated. Fortunately for the user, no special knowledge of control panel wiring is needed, since most computer centers have prewired panels. All that is normally required is to make certain that the proper panel is in place. Only in highly exceptional circumstances can the historian justify the expenditure of time in learning how to wire control panels.

Figure 5.11.    IBM 407 Tabulator

Courtesy of International Business Machines Corporation.

The basic IBM principle of reading cards is followed in tabulator operations. Electrical impulses are generated every time reading brushes and metal rollers make contact through holes in a card. These impulses then travel to the control panel which directs specific machine functions. Unlike the counter-sorter which reads only one column at a time, the tabulator simultaneously reads up to 80 columns and performs whatever functions the control panel calls for. Alphabetic and numeric characters are read with equal speed.

The transfer of information on punched cards to printed form is very useful. A permanent printed record of a set of data cards can be prepared quickly. This record can also be used to compare punched data with the original information for accuracy. Other uses of the tabulator might include a listing after each run on the counter-sorter or a listing of computer output in punched card form.

The Computer

Up to this point discussion has focused on the basic electromechanical card-processing machines likely to be found in any university computer center. In the remainder of this chapter the emphasis centers on the computer, what it is, what it does, and how it performs its operations. This section is not an introduction to computers. Rather it presents in simple and understandable language information that the writers feel is useful for an elementary computer orientation. For detailed information the interested reader will find a number of available works on the subject.[10]

A computer is a machine capable of performing mathematical and/or logical functions repetitively with phenomenal speed and accuracy. There are two types of computers, the analog and digital computer. The analog computer works with voltages or currents which represent quantities such as speed, weight, pressure, and the like. It is an analog device because its electrical circuits are wired to express equations which

Figure 5.12. **Basic component of all computers**

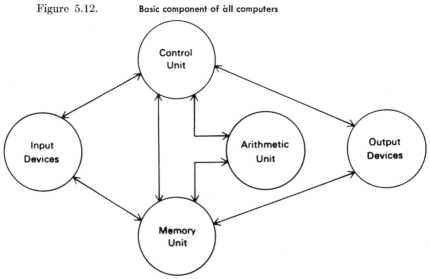

duplicate, say, operations of a mechanical system. The one to one relationship between the equations designated in the electrical circuits and the system being studied means that the movements of current symbolize the behavior of the corresponding system. Because of this characteristic the analog computer is used principally in engineering.

Figure 5.13. **IBM 2540 Card Reader**

The digital or general purpose computer handles data in the form of long strings of digits (zero and one) which can carry unlimited varieties of information. This gives the digital computer a flexibility the analog computer lacks. Also the digital computer lends itself to the kinds of analysis undertaken in the social sciences. Hereafter, when the word computer is used, it specifically means a digital computer.

The internal numbering system of digital computers is somewhat unusual. Instead of following the conventional ten-digit decimal system, the computer has a binary numbering system that uses only two digits—0 and 1—to represent alpha-numeric characters. Consideration of the technical aspects of binary numbers is unnecessary in this kind of book. It is enough to note that a binary system works well with the computer's

Figure 5.14.　　**IBM 1403 Printer**

electronic circuits since the 0 and 1 digits correspond respectively to the absence and presence of the current. It is highly unlikely that the historian will need to know any more than that the internal numbering system of a computer is binary and combines 0's and 1's to represent alphabetic characters, numeric characters of single or aggregate digits, and special characters.

Although digital computers vary in size and shape, they perform the same basic operations and require essentially the same kinds of equipment. Each computer must be able to carry out five operations: (1) read in instructions and data; (2) remember the program and data being used; (3) perform various mathematical or logical functions; (4) print out the results; (5) control the entire operation from start to finish. These five functions comprise the component units of a computer. Their interrelationship is diagrammed in Figure 5.12.

Input Unit

Before the computer can begin its work, the instructions or the program and data must be read in. The input unit does this. When the familiar punched cards are used for input to the computer, a card-reader like that in Figure 5.13 is used. It can read cards at the rate of 800 per minute. Faster input is achieved with plastic tape coated with a thin layer of magnetic material. Alphabetic, numeric, and special characters are recorded by configurations of miniscule magnetized areas of the tape. Punched card data may be transferred to tape, or a machine similar to the keypunch may be used to record coded data directly from data sheets to tape.

Tapes offer several advantages over punched cards. The same reel of tape can be reused by erasing old data and recording new data. More important, however, is the immense storage capacity of tapes. One-half inch of tape is required to record the contents of an 80 column card. The usual 12 inch tape reel of 2400 feet can store information that would require about 25,000 cards. In addition to compact storage, tape can be read much faster than cards. While the card-reader can process 800 cards per minute, a high speed tape drive can read the equivalent of 20,000 cards in the same length of time. This difference in input time is important since the internal operations of the larger computers are measured in microseconds. To rely exclusively on card-readers for input limits maximum utilization of computer speed, since the computer must wait for the instructions and data to be read in before processing a job. To reduce this inefficient use, many computer centers with heavy usage record input for a batch of jobs on tape via an auxiliary card-to-tape reader. The computer then reads directly from the tape, thus reducing the amount of time required to process each job. The term "off-line" designates input written on tape off the computer via an auxiliary unit which later is transferred to the computer. "On-line" refers to an input unit connected directly to the computer.

Memory Unit

Information becomes accessible to the computer only after it is entered into the storage unit, the computer's memory. This unit is not an electronic brain; rather it consists of areas that can be magnetized to represent information. The most common computer memory is called a core memory because it is composed of thousands of tiny magnetic rings through which wires about the size of a human hair pass. An electric current passing through the wires magnetizes each ring in one of two directions while another circuit keeps track of the direction of magnetism for each ring. Each ring accordingly contains one bit of binary information. Through a combination of magnetized cores in either direction each item of data is represented by a computer word. Core storage has been appropriately compared to a filing cabinet in which each item has a specific location with an "address" which serves to remind the computer where each digit or alphabetic character is stored.

One indication of computer capability is the capacity of its storage unit to accommodate computer words. This capacity is measured in units of 1000 words and is denoted by the letter K. For example, a computer with 16,000 words of storage has a memory capacity of 16K. Most of the IBM 7040 and 7090 computer series have 32K storage while computers in the new 360 series have 262K storage. A second generally accepted criterion of computer capability is the speed with which information may be called from storage and returned. This is called memory speed and usually is measured in microseconds. A core memory is designated a random access memory because each storage locator is as easy to reach as any other.

The usable storage of a computer may be effectively expanded beyond its core size by the use of auxiliary storage units which consist of magnetic tape and magnetic disk units. Tape units are commonly used to hold data temporarily while processing a job. Access time is slow compared to core memory since the information is stored sequentially and the tape must be moved each time to pick up the required data. Disk storage consists of what looks like a stack of phonograph records that rotate at high speeds. A recording head can read or write almost a million characters on each disk. The speed at which disks rotate permits faster access time than tape storage since less time is required for the recording head to position over the right part of the disk.

Arithmetic and Logical Unit

The actual processing of data is handled by a unit that performs mathematical calculations and makes logical comparisons. The arithmetic functions consist of addition, subtraction, multiplication, and division. Computer logic involves following a series of operations in a prescribed sequence and making decisions about whether a number is greater than, less than, or equal to another number. Tests of logical relations between two numbers permit the computer to decide which one of several prescribed alternatives should be followed at given points in the operation. While the arithmetic and logical functions are rather simple, when skillfully combined they greatly enhance the versatility of a computer to perform complex operations.

Output Unit

Once the specified operations in the arithmetic and logical unit are completed, the results are fed to the output section which usually consists of a high-speed printer, magnetic tape unit, and a card punch. The high speed printer, which is pictured in Figure 5.14, produces a printed page much like a typewriter does. However, instead of printing a character at a time, a chain driven mechanism prints an entire line at a time that contains 132 positions. In principle the high speed printer is like the tabulator described earlier except that it can print at the rate of 800 lines per minute. Since even this high rate of output is slow compared to internal computer speed, output frequently is recorded on magnetic tape and later converted to printed form on a tape-to-printer unit. This is known as "off-line" output and is employed where computer usage is heavy. A third output device is the card punch which produces punched cards containing the processing results in machine readable language. Punched information can also be printed, but its chief advantage is a permanent record of output that can be preserved or used as input data for another job.

There are other special-purpose output devices which are available at some computer centers. One is the X-Y plotter which draws graphs and maps under control of the computer. Another piece of output equipment is the cathode-ray tube which resembles a television screen. This unit permits immediate visual analysis of output without the delay of printing. Any part of the output on the screen can be photocopied as a permanent record. New types of output devices still in the experimental stages of development will add a fascinating dimension to this aspect of computer technology.

Control Unit

The heart of a computer is the control unit which directs all the computer operations. This unit receives instructions and in turn issues the necessary machine commands to execute the operation. The control unit also coordinates all computer functions and calls each unit into use at the proper time. In addition, it serves as a kind of a clearinghouse for every computer function and keeps track of what is being done. In summary the control unit insures that instructions are properly executed by the various component computer units.

The five component units previously mentioned—input, memory, arithmetic, control, and output—comprise computer hardware, a term that designates the physical apparatus of a computer. The rest of a computer system is made up of computer software. Software refers to equipment for preparing instructions that tell the computer what to do. Since a computer can only follow instructions, software is very important and requires detailed treatment.

Program Languages

Internal computer operations are based on a binary number system (consisting of 0's and 1's) called machine language that generates the electrical impulses necessary to run the computer. Since a computer "understands" only machine language, it is necessary that each instruction be in binary form. For example, an instruction to divide might be expressed by 01001000100000000000000000000000000. To write a program or complete set of instructions in machine language to process even a simple job is complex, to say the least. And the complexity increases as the computer operations become more intricate.

To obviate this difficulty programming languages have been developed. A program language consists of instructions in a specified manner not in binary form which are translated into the necessary machine language by a compiler program. Program languages and compiler programs comprise computer software and are absolutely essential for effective use of computer hardware.

Program languages may be classified as symbolic or problem oriented. In symbolic programming instructions are represented by such symbols as SUB for subtract, DIV for divide, and MUL for multiply. A program compiler generates one machine language for each symbol. The original program written in symbolic abbreviations is called a source program and the resultant machine language program is called the object program. While symbolic programming is widely used, it has several disadvantages. Each type of computer that uses symbolic programming has its own unique set of symbols. Hence, a symbolic language program can be used only on the computer for which it was written. Also, since each symbol generates only one machine language instruction, every step must be carefully spelled out. This means that the person writing the program must keep track of what each instruction does to avoid a programming error.

Problem oriented languages such as Fortran or Cobol are called higher level languages because they are more similar to English and more oriented to mathematics than symbolic languages. Once the fundamentals of a problem oriented language are mastered, it is relatively easy to use and offers a number of advantages. With a problem oriented language it is not necessary to be concerned with the internal computer operations that a particular program requires. It relieves the programmer of specifying each specific operation, since a single higher level instruction may generate a number of internal operations. This results in the computer handling details where programming errors are most likely to occur. Another advantage is that even though each different type of computer has its own compiler, a source program written in a higher level language can be translated into many different machine languages for different computers. Problem oriented languages save a programmer from bogging down in messy details and permit him to concentrate on the more important job of conceptualizing a problem and the best way to solve it.

A number of higher level languages have been developed, with Cobol and Fortran being the most widely used ones. Cobol, which stands for Common Business Oriented Language, is used primarily in accounting and other commercial operations. Fortran, which stands for Formula Translation, is used principally in the sciences. IBM developed Fortran in 1957 as an intermediate stage between human language and machine language since a program written in it can be compiled on any of the modern computers. Since 1957 Fortran has been updated several times, with the most advanced version being Fortran IV in 1961.

Another new programming language which is designed for use on the IBM 360 series computers is PL/1. Programming language 1 permits greater use of English in programming than does Fortran and bridges the gap between Fortran and Cobol. Since PL/1 does not rely exclusively on algebraic notations as Fortran does, the new programming language is highly compatible with the humanities. It appears that in the future PL/1 will eventually replace Fortran in computer oriented humanistic research.

Flowcharts

Before a computer program is written, the problem must be carefully analyzed and a diagram drawn that shows each major step of the operation. This diagram is called a flowchart. While programs may be written without first working up such a diagram, a flowchart is useful because it requires that the problem be defined and shows the logic of the various operations. The chart in Figure 5.15 is a humorous example of flowcharting. The explicit notation of the steps and logical sequences followed in getting up in the morning illustrates the fundamentals of diagramming a flowchart. Note the symbols used to identify interrelationships. The rectangular boxes indicate any processing operation except a decision. The diamond shape boxes indicate decisions. The lines leaving the decision symbols are labeled with the decision result. Arrows denote the direction of flow or movement through the diagram.

The fundamentals illustrated in the above flowchart are followed in outlining the major steps in any computer program. Of course the flowchart is as simple or complex as the program is. Figure 5.16 shows the sequences a computer follows in a rather simple program that analyzes election returns at the precinct level. In the example the 1932 presidential election returns for Tulsa, Oklahoma have been recorded on punch cards as discussed earlier in this chapter. A single card contains the following information for each precinct: year of election, county name, type of election, precinct number, and votes cast for the Democratic and Republican candidates. The object of the computer analysis is to determine for each precinct the total vote each party received, the percentage plurality between winning and losing parties, and which party won. In the figure below step 1 calls for the computer to read punched data from a card for each precinct. Step 2 computes the total vote cast for the two parties. In steps 3, 4, and 5 party percentages and a plurality percentage difference are calculated. Steps 6, 7, and 8 identify the winning party or a tie vote. Step 9 instructs the computer to print out the results for each precinct.

On the basis of the preceding flowchart a program was written in Fortran IV for an IBM 7040 computer. The program was written on

Figure 5.15.   "How to get up in the morning"

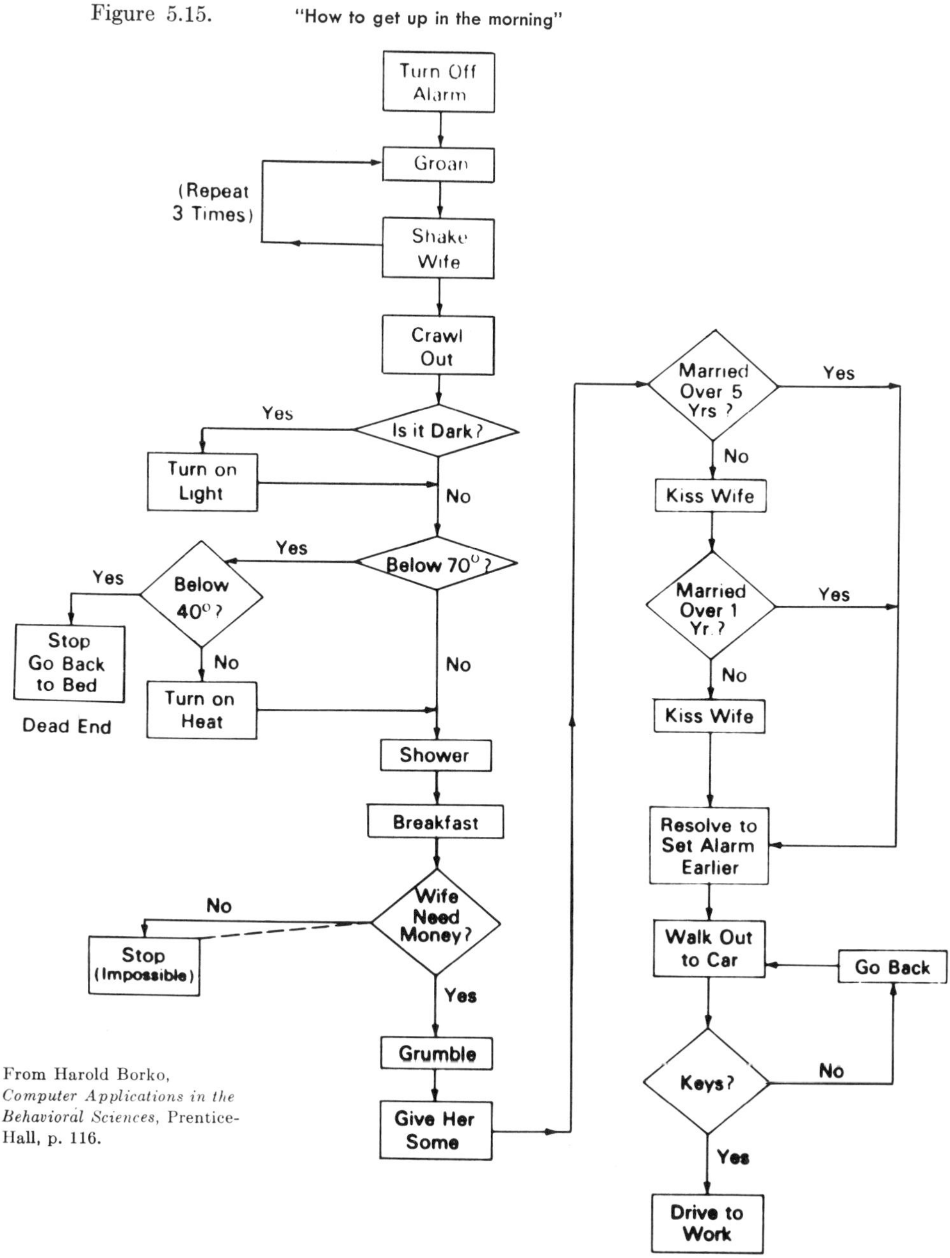

From Harold Borko,
*Computer Applications in the
Behavioral Sciences*, Prentice-
Hall, p. 116.

Figure 5.16   **Program Flowchart for Computing Election Percentages by Precinct**

STEPS   OPERATIONS

1   Read a card containing election year, county code, type of election, precinct number, Democratic and Republican vote

2   Compute the total vote cast for both parties

3   Compute Democratic percentage of vote. Democratic percentage equals Democratic vote divided by total vote

4   Compute Republican percentage of vote. Republican percentage equals Republican vote divided by total vote

5   Compute plurality percentage difference between winning party and losing party. Plurality percentage equals Republican percentage divided by Democratic percentage

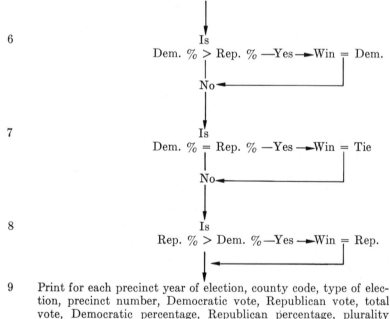

6   Is
     Dem. % > Rep. % —Yes—►Win = Dem.
     No◄

7   Is
     Dem. % = Rep. % —Yes—►Win = Tie
     No◄

8   Is
     Rep. % > Dem. %—Yes—►Win = Rep.

9   Print for each precinct year of election, county code, type of election, precinct number, Democratic vote, Republican vote, total vote, Democratic percentage, Republican percentage, plurality percentage, and winner

a standard coding form which was followed in punching the program into cards. These punched cards make up the source program deck which Fortran translates into an object program containing all the minute instructions in machine language (binary form) required to execute the program. The information on the 36 punched cards in the source program deck is shown in Figure 5.17. The function of each instruction on each card is described below.

Figure 5.17. Sample source program deck

```
STATEMENT                    INSTRUCTION                              CARD
NUMBER                                                                NUMBER

C     THIS PROGRAM COMPUTES PERCENTAGES AT THE PRECINCT LEVEL.         1
      DIMENSION YR(150),CTYCD(150),ELECT(150),PRCT(150),DEMVOT(150),   2
      1REPVOT(150),TOTVOT(150),DEMPCT(150),REPPCT(150),PLURAL(150),    3
      2WIN(150)                                                        4
C     VARIABLE NAMES FOR PRECINCT ANALYSIS.  YR=YEAR,CTYCD=COUNTY CODE,5
C     ELECT=ELECTION TYPE,PRCT=PRECINCT NUMBER, DEMVOT=DEMOCRATIC VOTE,6
C     REPVOT=REPUBLICAN VOTE, TOTVOT=TOTAL VOTE, DEMPCT=DEMOCRATIC PER-7
C     TAGE OF VOTE, REPPCT=REPUBLICAN PERCENTAGE OF VOTE, PLURAL=PERCEN-8
C     TAGE DIFFERENCE BETWEEN WINNING PARTY AND LOSING PARTY, WIN=WINNER9
      INTEGER WIN                                                      10
      READ (5,2)K                                                      11
  2   FORMAT (I3)                                                      12
      DO4 J=1,K                                                        13
  4   READ (5,6) CTYCD(J),ELECT(J),YR(J),PRCT(J),DEMVOT(J),REPVOT(J)   14
  6   FORMAT  (4X,I2,1X,I1,I3,34X,I3,3X,F5.0,3X,F5.0)                  15
      DO8 J=1,K                                                        16
      TOTVOT(J)=DEMVOT(J)+REPVOT(J)                                    17
      DEMPCT(J)=DEMVOT(J)/TOTVOT(J)*100.0                              18
      REPPCT(J)=REPVOT(J)/TOTVOT(J)*100.0                              19
      PLURAL(J)=ABS(REPPCT(J)-(DEMPCT(J))                              20
      IF(DEMVOT(J).GT.REPVOT(J))WIN(J)=1                               21
      IF(DEMVOT(J).EQ.REPVOT(J))WIN(J)=2                               22
      IF(REPVOT(J).GT.DEMVOT(J))WIN(J)=3                               23
  8   CONTINUE                                                         24
 10   WRITE (6,12)                                                     25
 12   FORMAT(1H1,4HYEAR,6X,5HCTYCD,6X, 5HELECT,6X,4HPRCT,6X,6HDEMVOT,  26
      16X,6HREPVOT,6X,6HTOTVOT,6X,6HDEMPCT,6X,6HREPPCT,6X,6HPLURAL,6X, 27
      23HWIN//)                                                        28
      DO14J=1,K                                                        29
 16   WRITE(6,18)  YR(J), CTYCD(J), ELECT(J), PRCT(J), DEMVOT(J),      30
      1REPVOT(J), DEMPCT(J), REPPCT(J), PLURAL(J), WIN(J)              31
 18   FORMAT (1X,I3,7X,I2,10X,I1,8X,I3,8X,F4.0,8X,F4.0,8X,F4.1,        32
      18X,F4.1,8X,I1)                                                  33
 14   CONTINUE                                                         34
      STOP                                                             35
      END                                                              36
```

```
C   THIS PROGRAM COMPUTES PERCENTAGES AT THE PRECINCT LEVEL.           1
```

This is a comment card that describes what the program does.
Comment cards, which the computer ignores, are for the benefit of the person using the program.

```
      DIMENSION YR(150),CTYCD(150),ELECT(150),PRCT(150),DEMVOT(150),      2
     1REPVOT(150),TOTVOT(150),DFMPCT(150),REPPCT(150),PLURAL(150),        3
     2WIN(150)                                                            4
```

These cards comprise a DIMENSION statement that identifies each variable used in the program and notifies the computer how much storage is to be allocated to each variable. Since there were 105 city precincts in the 1928 election, the 150's in parentheses actually provide more storage space than will be used. Three cards are required to express the DIMENSION statement. The 1 preceding the comma at the beginning of line 2 and the 2 preceding WIN at the beginning of line 3 notify the computer that they are continuation cards that belong to the DIMENSION statement.

```
C     VARIABLF NAMES FOR PRECINCT ANALYSIS.   YR=YEAR,CTYCD=COUNTY CODE,  5
C     ELECT=ELECTION TYPE,PRCT=PRECINCT NUMBER, DEMVOT=DEMOCRATIC VOTE,   6
C     REPVOT=REPUBLICAN VOTE, TOTVOT=TOTAL VOTE, DEMPCT=DEMOCRATIC PER-   7
C     TAGE OF VOTE, REPPCT=REPUBLICAN PERCENTAGE OF VOTE, PLURAL=PERCEN-  8
C     TAGE DIFFERENCE BETWEEN WINNING PARTY AND LOSING PARTY, WIN=WINNER  9
```

These are comment cards that identify each variable in the DIMENSION statement. Abbreviations are used as variable names because Fortran permits no more than six characters for a variable name. Cards 5–9 contain information only for the benefit of the person using the program. The machine does not use these cards.

```
      INTEGER WIN                                                        10
```

This notifies the computer that variable WIN is an integer.

```
      READ (5,2)K                                                        11
```

This instruction is to read a control card named K which tells the computer how many data cards are to be expected. The 5 in parentheses means that this information is to be read by the card-reader, while the 2 specifies the format. In Fortran a format statement tells the computer which columns in a card are to be read. The number assigned to a format statement is an arbitrary one; it could be any other number, though the same format number cannot be used again in the program to designate another instruction.

```
    2 FORMAT (I3)                                                        12
```

This statement tells the computer that the value of K is an integer variable located in the first three columns of the card. The I stands for "integer" and the "3" for the number of columns in the field.

```
DO4 J=1,K                                                      13
```

This is a DO Loop instruction sequence which means that the computer is to repeat the operation specified in statement 4 J number of times. The value of J ranges from 1 to the value of K, which has been previously determined.

```
4   READ (5,6) CTYCD(J),ELECT(J),YR(J),PRCT(J),DEMVOT(J),REPVOT(J)        14
```

After reading card 13 the computer expects the next card to complete the specifications of the DO loop. This card notifies the computer to read the variables CTYCD, ELECT, YR, PRCT, DEMVOT, REPVOT J number of times. The 5 within parentheses instructs this particular computer to read from its card-reader while the 6 indicates the format to be followed.

```
6   FORMAT (4X,I2,1X,I1,I3,34X,I3,3X,F5.0,3X,F5.0)                        15
```

This statement notifies the computer which columns contain the data that correspond to the variables listed in card 14. The first item within parentheses, 4X, means that the information in columns 1–4 is to be skipped. The I indicates that columns 5–6 contain the county code number and it is to be treated as a real number. The next figure, 1X, instructs the computer to skip the information in column 7. The I1 and I3 that follow denote the columns which hold the data for the type of election and the year of the election. The type of election is in column 8 and the election year is in columns 9–11. The 34X that follows I3 indicates that columns 12–45 are to be skipped. The I3 notifies the computer to read the next three columns, 46–48, which contain the precinct number. The last four items in the statement mean that the computer is to skip columns 49–51, read columns 52–56, skip columns 57–59, and read columns 60–64. The figures F5.0 represent the Democratic and Republican vote respectively. The F tells the computer DEMVOT and REPVOT are to be considered floating-point numbers, that is, numbers with decimal points. The .0 means that there are no digits past the decimal point. The numbers DEMVOT and REPVOT contain no decimal, but since percentages are to be calculated during the program they must be handled as floating-point or decimal numbers.

```
DO8 J=1,K                                                      16
```

This is another DO Loop sequence that instructs the computer to perform the calculations indicated by cards 16–23 J number of times. Since J represents the number of precincts, the sequence of operations specified in cards 16–23 is to be repeated for each precinct.

```
TOTVOT(J)=DEMVOT(J)+REPVOT(J)                                    17
```

This is a statement within the DO Loop that equates a variable called TOTVOT with the sum of DEMVOT and REPVOT. The J enclosed within parentheses indicates the number of times (or precincts) TOTVOT is to be computed.

```
DEMPCT(J)=DEMVOT(J)/TOTVOT(J)*100.0                              18
```

This statement is also within the DO Loop and equates a variable called DEMPCT with the result obtained by dividing DEMVOT by TOTVOT and multiplying the quotient by 100.0 to convert the answer to percents. The slash mark and the asterisk are Fortran instructions that respectively indicate division and multiplication.

```
REPPCT(J)=REPVOT(J)/TOTVOT(J)*100.0                             19
```

This instruction calculates the percentage for a variable called REPPCT as discussed under card 18.

```
PLURAL(J)=ABS(REPPCT(J)-(DEMPCT(J))                             20
```

PLURAL is the variable name for the plurality percentage difference between the parties' percentage of the total vote. ABS is a Fortran operand that automatically cancels negative signs that would appear say, if the DEMPCT for a precinct exceeded the REPPCT for the same precinct. The J's within parentheses indicate that PLURAL is to be calculated J number of times.

```
IF(DEMVOT(J).GT.REPVOT(J))WIN(J)=1                              21
IF(DEMVOT(J).EQ.REPVOT(J))WIN(J)=2                              22
IF(REPVOT(J).GT.DEMVOT(J))WIN(J)=3                              23
```

Cards 21, 22, 23 call for logical comparisons between the DEMVOT and REPVOT in each precinct. If the DEMVOT is greater than the REPVOT for a precinct then the WIN code is set at 1. If DEMVOT is not greater than REPVOT then nothing happens and the computer proceeds to execute the next instruction on card 22. With this card if REPVOT is greater than DEMVOT, then the WIN code is set at 2 to indicate a Republican victory in that precinct. If REPVOT did not exceed DEMVOT, then the WIN code is ignored and the computer proceeds to execute the next instruction. If DEMVOT is neither greater than nor lesser than REPVOT, the only logical alternative is that there was a tie. In this event the WIN code is set at 3 to indicate a tie vote. These three IF statements are part of the preceding DO Loop and are repeated J number of times or once for each precinct.

8   CONTINUE   24

This is a dummy statement that causes no execution and simply satisfies a Fortran rule for DO Loop statements.

10   WRITE (6,12)   25

This tells the computer that the next operation is to write output. The 6 within parentheses indicates that its printer is to be used (the output could be written on tape or cards) and the 12 indicates the format to be followed. The printed information consists of headings across the top of a page.

12   FORMAT(1H1,4HYEAR,6X,5HCTYCD,6X,5HELECT,6X,4HPRCT,6X,6HDEMVOT,   26
     16X,6HREPVOT,6X,6HTOTVOT,6X,6HDEMPCT,6X,6HREPPCT,6X,6HPLURAL,6X,   27
     23HWIN//)   28

This statement instructs the computer what labels are to be printed out, how many printing positions each label has, and the number of blank spaces between each label. The 1H1 is a standard notation for the printer carriage mechanism that causes the printing to begin at the top of the page. Each label is preceded by a number and an H. The H means that the Hollerith method of representing alphabetic characters is to be used and the number preceding the H indicates the number of printing positions. The letters immediately following each H will be printed. For example, 4HYEAR means the four letters YEAR will be printed. Each number that is followed by an X (5X or 6X) represents the number of blank spaces between each label. This is done to make the printed output visually attractive and easy to read. The two slash marks at the end of the format statement instruct the printing carriage to skip a line between the label headings and the first line of output. Since this FORMAT statement required three cards, continuation numbers (the 1 and 2 below and to the left of FORMAT) notify the computer how many to expect.

     DO14J=1,K   29

This is a DO Loop instruction that tells the computer to repeat the operation specified in cards 30–31 J number of times.

```
16  WRITE(6,18)   YR(J), CTYCD(J), ELECT(J), PRCT(J), DEMVOT(J),        30
    1REPVOT(J), DEMPCT(J), REPPCT(J), PLURAL(J), WIN(J)                  31
```

This statement commands the computer to call from memory
the output for each variable listed and print it according to
FORMAT 18.

```
18  FORMAT (1X,I3,7X,I2,10X,I1,8X,I3,8X,F4.0,8X,F4.0,8X,F4.1,           32
    18X,F4.1,8X,I1)                                                      33
```

This FORMAT statement specifies the number of columns
each variable occupies and the spacing between each variable. The numbers preceded by I are integer numbers while
the numbers preceded by F are floating-point numbers.
Floating-point numbers are used because percentages are involved.

```
14  CONTINUE                                                            34
```

This instruction is the same as that in card 24.

```
    STOP                                                                35
    END                                                                 36
```

The STOP indicates that no more statements are to be executed in the program while the END is a signal to the compiler that the end of the program has been reached and that
production of the object program may be initiated.

Part of the output resulting from the use of this program with
the precinct data cards for the 1928 presidential election is given in
Figure 5.18. The "1" under ELECT is the code number for presidential
elections. The decimal point in the figures under DEMVOT, REPVOT,
and TOTVOT indicate they are floating-point numbers. Actually the
decimal means nothing and could be eliminated. To include these steps
in the example program would present unnecessary complications. The
computer required less than a minute to perform the calculations and
print out the results.

Debugging

Once a program is written, how does the user know it is correct
and will perform the expected operations? With a simple program like
the one just described, visual checking can disclose obvious syntax errors
or violations of Fortran rules. In addition, the compiler, which translates
the source program into machine language, has a built-in diagnostic
check for syntax errors. For example, if a variable is not defined or
a comma is not placed at the proper place in a statement, the compiler
will not execute the program. Instead, a list of errors will be printed
out so that the necessary corrections can be made. This operation is

generally known as "debugging" a program. Computer systems that use Fortran IV have a special diagnostic compiler system called WATFOR which very quickly compiles a program (usually in less than a minute for complex programs) without punching out the object (binary) deck. If WATFOR is unavailable then "debugging" may be done by submitting a job for processing as though the program had already been "debugged."

Figure 5.18.   **Computer print out of example program**

YEAR	CTYCD	ELECT	PRCT	DEMVOT	REPVOT	TOTVOT	DEMPCT	REPPCT	PLURAL	WIN
932	55	1	01	216.	154.	370.	58.4	41.6	16.8	1
932	55	1	02	148.	117.	265.	55.8	44.2	11.7	1
932	55	1	03	254.	171.	425.	59.8	40.2	19.5	1
932	55	1	04	235.	121.	356.	66.0	34.0	32.0	1
932	55	1	05	207.	200.	407.	50.9	49.1	1.7	1
932	55	1	06	235.	231.	466.	50.4	49.6	0.9	1
932	55	1	07	181.	99.	280.	64.6	35.4	29.3	1
932	55	1	08	129.	159.	288.	44.8	55.2	10.4	3
932	55	1	09	172.	156.	328.	52.4	47.6	4.9	1
932	55	1	11	209.	213.	422.	49.5	50.5	0.9	3
932	55	1	12	192.	165.	357.	53.8	46.2	7.6	1
932	55	1	13	109.	91.	200.	54.5	45.5	9.0	1
932	55	1	14	202.	136.	338.	59.8	40.2	19.5	1
932	55	1	15	157.	157.	314.	50.0	50.0	0.0	2
932	55	1	16	213.	159.	372.	57.3	42.7	14.5	1
932	55	1	17	133.	123.	256.	52.0	48.0	3.9	1
932	55	1	18	148.	132.	280.	52.9	47.1	5.7	1
932	55	1	19	227.	176.	403.	56.3	43.7	12.7	1
932	55	1	20	126.	87.	213.	59.2	40.8	18.3	1
932	55	1	21	202.	139.	341.	59.2	40.8	18.5	1
932	55	1	23	202.	201.	403.	50.1	49.9	0.2	1
932	55	1	24	162.	146.	308.	52.6	47.4	5.2	1
932	55	1	25	174.	174.	348.	50.0	50.0	0.0	2
932	55	1	27	231.	170.	401.	57.6	42.4	15.2	1
932	55	1	28	206.	133.	339.	60.8	39.2	21.5	1
932	55	1	29	130.	129.	259.	50.2	49.8	0.4	1
932	55	1	30	168.	111.	279.	60.2	39.8	20.4	1
932	55	1	31	257.	130.	387.	66.4	33.6	32.8	1
932	55	1	32	146.	69.	215.	67.9	32.1	35.8	1
932	55	1	33	152.	96.	248.	61.3	38.7	22.6	1
932	55	1	34	156.	122.	278.	56.1	43.9	12.2	1
932	55	1	35	302.	191.	493.	61.3	38.7	22.5	1
932	55	1	36	243.	157.	400.	60.7	39.2	21.5	1
932	55	1	37	138.	74.	212.	65.1	34.9	30.2	1
932	55	1	38	243.	157.	400.	60.7	39.2	21.5	1
932	55	1	39	184.	170.	354.	52.0	48.0	4.0	1
932	55	1	40	199.	163.	362.	55.0	45.0	9.9	1
932	55	1	41	178.	117.	295.	60.3	39.7	20.7	1
932	55	1	42	220.	165.	385.	57.1	42.9	14.3	1
932	55	1	43	177.	184.	361.	49.0	51.0	1.9	3
932	55	1	44	294.	189.	483.	60.9	39.1	21.7	1
932	55	1	45	223.	211.	434.	51.4	48.6	2.8	1
932	55	1	46	201.	149.	350.	57.4	42.6	14.9	1
932	55	1	47	84.	53.	141.	62.4	37.6	24.8	1
932	55	1	49	146.	120.	266.	54.9	45.1	9.8	1
932	55	1	50	229.	147.	376.	60.9	39.1	21.8	1
932	55	1	51	105.	117.	222.	47.3	52.7	5.4	3
932	55	1	52	106.	100.	206.	51.5	48.5	2.9	1
932	55	1	53	86.	104.	190.	45.3	54.7	9.5	3
932	55	1	55	174.	100.	274.	63.5	36.5	27.0	1
932	55	1	56	193.	159.	352.	54.8	45.2	9.7	1
932	55	1	58	234.	154.	388.	60.3	39.7	20.6	1
932	55	1	59	189.	129.	318.	59.4	40.6	18.9	1
932	55	1	60	216.	119.	335.	64.5	35.5	29.0	1

If programming errors are encountered, the compiler process will terminate and a list of errors will be printed out.

Nonprogrammers and Computer Use

Certainly an understanding of the mechanics of writing and "debugging" the sample program, which is quite simple and scarcely uses the computer's capabilities, does not qualify the historian as an expert programmer. But is it even necessary for historians to acquire programming expertise to use the computer effectively? While strong arguments can be marshalled to support the position that computer users should learn to write their own programs, this is unrealistic and impractical for most historians. Programming for computers is a profession that requires continual involvement and upgrading in expertise as newer computers and programming languages become available. Only the extraordinarily gifted historian can maintain proficiency in programming and his own discipline. Another reason why historians need not write their own programs is the increasing availability of standardized programs. Every computer center maintains a library of programs and writeups that are operational on the computer there. Since a writeup usually describes each program, programming experience is not a prerequisite for its use. Furthermore, information about new programs is available through SHARE (Society to Help Avoid Repetitive Effort) and in such publications as *Behavioral Science, Psychological Abstracts,* and *BMD: Biomedical Computer Programs.* Also IBM provides a general listing of programs for IBM equipment in its *Catalog for IBM Data Processing Systems.* Frequently, many of these programs can be adapted with little difficulty to fit the requirements of a specific research design.

The argument that historians do not have to become expert programmers in order to use the computer most emphatically *does not mean* that effective computer usage has no requirements. Indeed, while the computer offers tremendous advantages, it certainly imposes requirements on historians. Perhaps the most crucial one is that of working out a research design for computer analysis that is carefully planned and designed to test explicitly formulated hypotheses derived from systematic conceptualization of an important research problem. Indeed, this may prove to be the computer's greatest challenge to historians. Another requirement is that of acquiring a familiarity with the concepts and language of computer operations so historians can comprehend the problems and feasibility of using the computer. This familiarity assumes a working knowledge of the fundamentals of programming. While few historians have even this level of computer knowledge, it requires only a small investment of time. Most computer centers offer noncredit computer orientation courses that meet twice a week for six or eight weeks. Enrollment in such a course will amply repay historians with a greater appreciation of the capability of the computer as a research tool. Of course the greatest reward comes with the actual use of the computer. Repeated use of the computer develops the ability to utilize it to greater advantage in dealing with other questions that previously were not amenable to traditional methods of historical research.

Until programming languages become as easy to use as the English language, even historians with a working knowledge of the fundamentals of programming will require programming assistance. Such assistance may come from a graduate student with computer science training or a fellow faculty member in another discipline who programs. More likely, however, historians will turn to a professional programmer on the computer center staff. Most programmers are aware of their unique status as highly skilled professionals who function as virtually irreplaceable intermediaries between a very complex machine and relatively uninformed users. Consequently, professional programmers tend to live in a world of their own, which makes communication with them somewhat difficult. Furthermore, because of the demand for his skill a professional programmer tends to have little patience with the person who in effect dumps a research problem in his lap and expects him to solve it. Under ordinary circumstances a historian who expects total programming assistance will soon discover that the programmer has little time for him or his problem.

The obvious need for programming assistance suggests that historians must learn to work with programmers on their terms. This means that it is the historian's responsibility to define explicitly what he wants to do. No historian should expect a programmer, or anyone else for that matter, to understand what he wants to do unless he can explain it with clarity and precision himself. Another effective base for communication between a historian and a programmer is the former's familiarity with the concepts of computer operation and analysis so that he can speak and understand, however haltingly, the programmer's language. Finally, and by no means least, is the desirability of diagramming a flowchart of the major steps of the proposed computer operations. If historians are willing to work within this framework, they will find most professional programmers very cooperative and even interested in their projects.

A note of warning is an appropriate conclusion for this section on the computer. Even though a program is correctly written with the assistance of a professional programmer, it is not a panacea for fundamental errors that may be associated with the method used to solve a problem. The results of computer analysis are no better than the method and data fed into the computer. Apropos to this point and well worth remembering is a fundamental axiom of computer applications: "Garbage in, garbage out." A correctly executed program can never overcome the deficiencies of a faulty research design or transform a trivial research problem into an important one. The responsibility for the research design and the importance of the research problem quite properly rest with the historian, not the programmer.

This chapter has been devoted chiefly to a consideration of the fundamentals of data processing at two levels—electromechanical and electronic data processing. The emphasis has been that of providing an orientation for historians so that they may intelligently consider the problems, relevance, and feasibility of employing data processing in their research. The following chapter will offer concrete examples of application of data processing techniques in historical research.

FOOTNOTES

1. This chapter draws heavily upon the insights and information in the following works: Harold Borko, ed., *Computer Applications in the Behavioral Sciences* (Englewood Cliffs: 1962); Kenneth Janda, *Data Processing: Applications to Political Research* (Evanston: 1969); James A. Saxon and W. W. Steyer, *Basic Principles of Data Processing* (Englewood Cliffs: 1967); Peter A. Stark, *Digital Computer Programming* (New York: 1967); Donald J. Veldman, *Fortran Programming for the Behavioral Sciences* (New York: 1967). In addition the following IBM reference manuals were very helpful: "An Introduction to IBM Punched Card Data Processing," "IBM 26 Printing Card Punch," "IBM 82, 83 and 84 Sorters," "IBM 402, 403, and 419 Accounting Machines," and "IBM 548, 552 Interpreters."

2. Data also can be converted into machine readable language by means of punched paper tape, magnetic tape, an optic character reader, or a magnetic ink character reader. For information on these devices the reader should consult Kenneth Janda, *Data Processing.*

3. Certain kinds of electromechanical accounting equipment can multiply and divide, but usually they are not available at a university computer center.

4. BORKO, *Computer Applications in the Behavioral Sciences,* p. 60.

5. RALPH BISCO, "Policies and Standards for Coding Data," (Mimeographed, Inter-University Consortium for Political Research), p. 5.

6. BISCO, "Policies and Standards for Coding Data," p. 2.

7. BISCO, "Policies and Standards for Coding Data," p. 2.

8. For specific information about standardized codes write the Director of Data Recovery, Inter-University Consortium for Political Research, Box 1248, Ann Arbor, Michigan.

9. BISCO, "Policies and Standards for Coding Data," p. 8.

10. See Chapter 7, Part 3, Section A5.

COMPUTER APPLICATIONS IN HISTORICAL RESEARCH

CHAPTER SIX

Classification is a necessary first step for all historical explanation. The process of selecting certain information as relevant relies upon classification schemes based on explicit or implicit theory. An implicit classification framework is evident in the sequence in which information is arranged. An explicit classification framework emerges whenever events, institutions, processes, or individual behavior patterns are depicted as representative of general types. For example, the description of a politician as liberal, radical, or conservative, or of a family as upwardly mobile or downwardly mobile, or a belief system as progressive or reactionary indicates the existence of a classification scheme. What T. H. Huxley wrote almost a century ago with regard to biological classification holds true for historical classification.

> By classification of any series of objects is meant the actual, or ideal arrangement together of those which are alike, and the separation of those which are unalike; the purpose of this arrangement being to facilitate the operations of the mind in clearly conceiving and retaining in the memory, the characteristics of the objects in question.[1]

Equally true is the fact that classification requires measurement of the phenomena under consideration to determine the extent to which certain properties are present or absent. Unfortunately, the classification techniques of many historians tend to be highly impressionistic. Usually a hunch or insight prompts an investigation of phenomena which intuitively seem related. Out of this general impression of the data several "dominant" characteristics are selected as criteria by which similarity or dissimilarity are determined. Such an approach suffers from obvious deficiencies, one of which is that the human mind can not make a comprehensive comparison of the characteristics of the objects under consideration. A second deficiency is the imprecision in assessment of relationships. Vagueness is concealed in such terms as "typical," "usually," "strong," "weak," "more," "less," "increasing," and many other such words that are part of the historian's working vocabulary.

Quantitative analysis as described in a previous chapter emphasizes systematic analysis, empirical verification, and mathematical measurement, and thereby offers historians a way to obviate some of the problems of impressionist classification encountered in dealing with certain kinds of questions in historical research. The linking of mathematical measurement with the computational capability of the computer results in a tremendous saving of a researcher's time and energy. In addition, a computer can keep track simultaneously of the interrelationships among a large number of variables, something which the human mind does very poorly.

Simple Data Analysis

The capacity of computers for rapid processing and ordering of large bodies of information should be of basic assistance in helping historians organize bodies of relevant but disordered information and at the same time search systematically for meaningful patterns. Historians who exploit this capability of a computer expand enormously their own capacity so that all available and relevant data can be used. Furthermore, greater confidence is warranted in conclusions based on a systematic analysis of all relevant data.

The Court of Wards, 1624–1635

Roy E. Schreiber's computerized study of the Court of Wards, 1624–1635 offers an illustration of simple computer data analysis in historical research.[2] For historians of Stuart England interested in the conflicts between Parliament and James II over royal finance, the activities of the Court of Wards are of special interest because they provided the greatest single source of non-Parliamentary revenue for the king. In order to evaluate the operations of the Court in the early Stuart years, Schreiber examined some 2000 Wardship cases settled between 1622 and 1637. Initially, two major questions stood out. Were the purchase arrangements for Wardships the same when the buyer was related to the Ward vis-à-vis a buyer unrelated to the Ward? Who were the unrelated buyers? Did they share any traits in common? In particular, were the "unrelated buyers" primarily royal servants who received Wardships as a kind of supplement to their salaries and fees "as was in the case until 1610?"[3]

Schreiber elected to use a computer to help answer these questions. Raw data, consisting of the assessed value of each Wardship, its purchase price, the county in which the property was located, the name and relationship of the purchaser to the Ward, and the college that the purchaser attended, for 2000 Wardships were recorded on punch cards. The computer was first instructed to calculate the number and percentage of related purchasers during four time periods: 1622–1623 (Cranfield); 1624–1628 (Early Naunton); 1629–1635 (Later Naunton); 1636–1637 (Cottington). Calculations were made for all of England and Wales as well as for each county. The output for five counties and all of England during each of the four time periods is shown in Table 6.1. The column headings identify each purchaser group. Of special interest are the ones labeled "Unknown" and "None" which denote purchasers unrelated to the Ward.

The second phase of the program analyzed the relationship between the price of Wardships and the value of the Ward's property. The analysis was conducted so as to yield correlation coefficients for each county as well as England and Wales during each of the four time periods. The data on which the correlation coefficients are based were the rent value at the time of death and the feodary certificate evaluations. Schreiber's analysis revealed two periods—Cranfield and Later Naunton—in which the correlation between the price of Wardships and feodary evaluation was low, .589 and .636 respectively. Examination of the calculations for related purchasers revealed that during Cranfield's tenure, 33 percent of the purchasers had no connection with the Wards. During Naunton's later years, 42 percent of the purchasers had no connection. The conclusion is that "the greater the number of relatives buying Wardships, the more closely the feodary certificates were followed."[4]

The tendency to ignore feodary certificate evaluation during the period when 42 percent of the purchasers had no known connection with the Ward purchased suggested that an analysis of this group should be made. This disclosed that many of the group were multiple purchasers.

TABLE 6.1 Wardship Purchaser Patterns From Roy E. Schreiber, "Studies in the Court of Wards, 1624–1635—The Computer as an Aid," *Computer Studies in the Humanities and Verbal Behavior,* (January 1968), pp. 36–42. Reprinted by permission of the author.

CRANFIELD	WARD	MOTHER	MOTHER & OTHER	GRANDF.	GRANDM.	UNCLE	COUSIN	OTHER	UN-KNOWN	NONE	TOTAL
Total	11	119	24	4	4	11	0	27	87	8	294
	4%	40%	8%	1%	1%	4%	0	9%	30%	3%	100%
London & Middlesex	1	7	2	0	0	2	0	5	4	0	21
	5	33	10	0	0	10	0	24	19	0	100
Kent	0	4	1	0	0	1	0	3	6	0	15
	0	27	7	0	0	7	0	20	40	0	100
Essex	2	10	1	0	1	2	0	3	5	0	24
	8	42	4	0	4	8	0	13	21	0	100
Gloucester	1	14	3	0	0	2	0	0	3	0	23
	4	61	13	0	0	9	0	0	13	0	100
Yorkshire	0	11	2	1	1	2	0	1	17	5	40
	0	27	5	2	2	5	0	2	42	12	100
EARLY NAUNTON											
Total	6	138	65	10	6	23	2	19	64	2	335
	2%	41%	19%	3%	2%	7%	1%	6%	19%	1%	100%
London & Middlesex	0	7	6	1	0	3	0	1	4	0	22
	0	32	27	5	0	14	0	5	18	0	100
Essex	0	11	6	0	1	0	0	2	12	0	32
	0	34	19	0	3	0	0	6	38	0	100
Somerset	0	14	4	1	0	1	0	0	4	0	24
	0	58	17	4	0	4	0	0	17	0	100
Gloucester	1	11	5	0	0	1	0	0	6	0	24
	4	46	21	0	0	4	0	0	25	0	100
Yorkshire	0	18	7	1	3	6	0	3	3	1	42
	0	43	17	2	7	14	0	7	7	2	100

TABLE 6.1 (Continued)

LATER NAUNTON

											Total
Total	25	289	23	6	1	9	0	81	285	41	760
	3%	38%	3%	1%	0%	1%	0%	11%	37%	5%	100%
Essex	1	10	2	0	0	2	0	5	13	13	46
	2	22	4	0	0	4	0	11	28	28	100
Norfolk	0	10	2	0	1	0	0	3	11	5	32
	0	31	6	0	3	0	0	9	34	16	100
Lincoln	1	13	1	0	0	0	0	11	22	2	50
	2	26	2	0	0	0	0	22	44	4	100
Yorkshire	2	36	4	2	0	1	0	14	62	4	125
	2	29	3	2	0	1	0	11	50	3	100

COTTINGTON

											Total
Total	8	133	3	0	2	0	0	20	95	15	276
	3%	48%	1%	0%	1%	0%	0%	7%	34%	5%	100%
Essex	1	3	1	0	0	0	0	3	4	4	16
	6	19	6	0	0	0	0	19	25	25	100
Somerset	1	8	0	0	1	0	0	0	1	1	12
	8	67	0	0	8	0	0	0	8	8	100
Wales	0	11	0	0	0	0	0	2	8	0	21
	0	52	0	0	0	0	0	10	38	0	100
Derby	0	7	0	0	0	0	0	6	1	0	14
	0	50	0	0	0	0	0	43	7	0	100
Yorkshire	2	26	0	0	1	0	0	3	25	0	57
	4	46	0	0	2	0	0	5	44	0	100

There was no identifiable pattern of university education or legal training among the group. The analysis did reveal, however, that many of the multiple purchasers were connected with the Court of Wards itself or with someone on it and lived in London or its suburbs.

TABLE 6.2. Occupational Distribution in 1880 and 1890

BOSTON MOBILITY STUDY   1880 SAMPLE   15 APRIL 1966

CONTINGENCY TABLE NO. 48

SUB-TABLE OF UNITS WITH PROT   UNKNON ON VAR   6   RELIGIOUS GROUP

VAR 24 1890 OCCUPATION (rows) × VAR 20 1880 OCCUPATION (columns)

1890 \ 1880		NO WRK UND 20	PROFES SIONAL	PROPTR OFFICL	SEMI-PROFES	CLERCL SALES	PETTY PROPTR	SKILED WORKMN	SMISKL SERVC	UNSKIL MENIAL	NO OCC	TOTAL
NO WRK UND 20	%	34.2					1.5		1.0			(11.3)
	N	113					1		1			115
PROFES SIONAL	%	2.1	83.3	2.1		1.4	1.5	1.1	1.0		20.0	(3.8)
	N	7	20	2		2	1	2	1		4	39
PROPTR OFFICL	%	2.1	4.2	81.9		10.9	7.5	1.1		2.4	5.0	(10.8)
	N	7	1	77		16	5	2		1	1	110
SEMI-PROFES	%	2.4			85.7							(1.4)
	N	8			6							14
CLERCL SALES	%	33.3		4.3		65.3	11.9	4.3	7.8	7.1	15.0	(23.6)
	N	110		4		96	8	8	8	3	3	240
PETTY PROPTR	%	0.6	4.2			1.4	59.7	2.2	6.9	4.8	5.0	(5.8)
	N	2	1			2	40	4	7	2	1	59
SKILED WORKMN	%	10.6	4.2	1.1		6.8	4.5	81.7	11.8	19.0	10.0	(22.0)
	N	35	1	1		10	3	152	12	8	2	224
SMISKL SERVC	%	3.3		1.1		5.4	4.5	2.2	59.8	14.3	5.0	(9.3)
	N	11		1		8	3	4	61	6	1	95
UNSKIL MENIAL	%	2.1		1.1		0.7	1.5	2.2	4.9	47.6		(3.8)
	N	7		1		1	1	1	4	20		39
NO OCC	%	9.1	4.2	8.5	14.3	8.2	7.5	5.4	6.9	4.8	40.0	(8.2)
	N	30	1	8	1	12	5	10	7	2	8	84
TOTAL		330	24	94	7	147	67	186	102	42	20	1019
PERCENT		(32.4)	(2.4)	(9.2)	(0.7)	(14.4)	(6.6)	(18.3)	(10.0)	(4.1)	(2.0)	(100.0)

Reprinted by permission of Stephan Thernstrom.

Schreiber concludes that his computer analysis yielded the following: (1) During Naunton's early years most purchasers of Wardships were friends and relatives of the Wards; (2) during Naunton's later years a group of multiple purchasers with no connection with the Wards began to purchase heavily in the Wardships; (3) these multiple purchasers were associated with the Court of Wards or with one of its members and tended to live in or near London; and (4) most multiple purchasers paid less than the feodary evaluation for the Wardships they purchased. "The pattern which emerges for Naunton's mastership is the reverse of Elizabethan times. Wardships were no longer going to supplement the income of Royal appointees. Instead they were going either to relations of the Ward or to professional Wardship hunters with the clear aim of raising revenue for the Crown."[5]

In evaluating the computer's contribution to this study, Schreiber notes that while only 45 minutes were required for the analysis, preparation of the data for the run required 80 hours. Yet this is a modest investment of time considering the fact that the computer uncovered relationships which in all probability would never have occurred to him. In this regard Schreiber concludes that while his computer-assisted analysis answered some questions, it raised some new ones. And these new questions are probably of greater consequence than the ones which prompted the initial research.

Social Mobility in Boston, 1880–1963

Stephen Thernstrom uses a more sophisticated method in his study of the social structure of Boston from 1880 to 1963.[6] The data consist of three random samples of information about males in Boston. The first sample of 3730 males came from the 1880 manuscript schedules of the U.S. Census. The second sample of 2166 males came from marriage license applications in Boston in 1910. The third sample of 1722 males was obtained from the 1930 birth certificates of that year. In the latter two cases, the samples were of fathers and sons, since both the marriage license applications and birth certificates gave the father's name. The source of each sample provided information consisting of age, occupation, religious affiliation, and place of residence of each male adult. Supplementary information was obtained from city directories and tax books. Thus, the three samples make up three separate data files for three distinct points in time as well as movement across time.

Thernstrom's research goal was to identify mobility patterns of different ethnic groups, opportunities of social mobility during industrialization, and the relationship between residential mobility, property mobility, and occupational mobility. The principal assistance the computer renders is that of cross tabulating variables against each other. The computer program used is part of the package programs of the Data Text System developed at the Harvard Department of Social Relations for operation on an IBM 7094 computer. Table 6.2 is a sample page of the printout of this program. This particular contingency table cross tabulates the occupational distribution in 1880 with that of 1890. The researcher concerned with occupational mobility would be particularly interested in the entries under the first column heading, Number of Workers Under Twenty. In this instance, 33.3 percent of the sample in 1880 who were under 20 had entered clerical sales occupations by

1890. This statistic becomes more meaningful, however, when it is compared with tabulations of occupational distributions at other periods. The program generates similar contingency tables for different time periods so that trends of occupational mobility, along with other trends, are easily isolated. While Thernstrom's study is not yet completed, he himself points out that his study of Boston would be "an unmanageable task" without the assistance of a computer.[7]

Schreiber and Thernstrom used the computer to answer the basic questions: what happened? when? where? and who participated? In dealing with a political institution, Schreiber asked who got what? when? and how? The answer to "how" does not immediately spring from the printout, but the patterns in the tabulated results narrow the search for the answer. Note that the statistical manipulation of data drawn from many sources was possible only after the researcher had set up a classification system, such as Thernstrom's occupational categories, and had decided on a common method and measurement. The ordering of data from a variety of sources and their summarization in aggregate numbers or percentages enhances systematic explanation. Computerized cross tabulation of selected data items from a larger body of information can also isolate patterns of relationships which may not have been anticipated.

Multiple and Partial Correlation

Since Turner, historians have searched for a multiplicity of geographical, economic, and social variables that correlate highly with political behavior. Frequently, only two variables at a time are correlated in this search. Common sense tells us that in the real world there are few instances where other variables do not affect the relationship between any two given variables. Multivariate ("many-variables") statistical techniques are available to measure the unique contributions of each independent variable to a single outcome or dependent variable. These techniques simultaneously take into account a number of different variables and calculate the effect of all the independent variables on the dependent variable and then measure the relation of each independent variable to the dependent variable. One such technique is called multiple and partial correlation. The elaborate mathematical formulae of these coefficients are so laborious and time-consuming that only computer assistance makes their use feasible in historical research that involves considerable data. Fortunately, packaged computer programs are available that will run very efficiently on most second and third generation computers.

Thomas B. Alexander has illustrated the value of multiple and partial correlation analysis in his study of "The Basis of Alabama's Ante-Bellum Two-Party System."[8] Alexander and his associates tested the Turnerian theory that Alabama Whigs tended to be wealthy, aristocratic planters and merchants, while the Democrats were poorer farmers. In one part of the study they used computerized multiple and partial correlation analysis of demographic and election data to identify at the county level the relationship between Whig voting strength and selected indicators of economic development recorded in the 1850 and 1860 United States censuses. The hypothesis being tested is that if economic considerations underlay alignment with the Whig party one could expect multiple and partial correlation analysis to disclose a "significantly high positive relationship between percentages of Whig vote and prevalence of economic prosperity as shown by land values, production levels, slaveholding, or any of the clues to advanced economic development and affluence."[9]

The data were structured in terms of independent and dependent variables. The latter consisted of the percentage presidential vote by county of the Whig party between 1840 and 1852, the American party in 1856, and the Douglas, Breckinridge, Bell Whig, and Douglas Bell factions in 1860.[10] The independent variables were 11 indicators of economic development for each Alabama county in 1850 and 1860 as recorded in the United States censuses. In order to be comparable with the voting data, the census information was converted into percentages by locating the county with the highest value on each of the 11 indicators and expressing the information for each county as a percentage of the maximum figure.

These percentage calculations along with election returns for each county were recorded in a punch card, with the independent variables entered first and followed by the dependent variables. These cards formed the input for a computer program run on a UNIVAC 1107 at the University of Alabama.[11] Since this program yielded a number of calculations not germane to this particular study, only the multiple and partial correlation coefficients and the coefficient of determination were used. These coefficients were extracted from the printout and rearranged as in Table 6.3.

At the bottom of the table are listed the multiple correlation coefficients R and coefficients of determination R^2 among the 11 independent variables and each voting percentage. The fact that all the multiple correlations are rather high suggests that there was a close connection between voting behavior and the collective impact of the economic variables. However, the coefficients of determination, which estimate the extent of voting behavior that all these variables actually explain, indi-

TABLE 6.3 Coefficients of Multiple and Partial Correlation between Each of Ten Economic Characteristics Selected from the Census of 1850 or 1860 and Political Party Division of the Popular Vote in Presidential Elections from 1840 through 1860 for All of the Counties of Alabama

From Thomas B. Alexander, et. al., "The Basis of Alabama's Ante-Bellum Two-Party System," *The Alabama Review*, XIX (1966), pp. 243–276. Reprinted by permission of the University of Alabama Press.

ECONOMIC CHARACTERISTICS	ECONOMIC CHARACTERISTICS FROM THE CENSUS OF 1850 PARTIAL CORRELATION WITH:				AMERICAN PARTY	ECONOMIC CHARACTERISTICS FROM THE CENSUS OF 1860 PARTIAL CORRELATION WITH:			
	WHIG % 1840	WHIG % 1844	WHIG % 1848	WHIG % 1852	% 1856	DOUGLAS FACTION % 1860	BRECKIN-RIDGE % 1860	BELL WHIG % 1860	DOUGLAS AND BELL % 1860
Cotton production	.09	.11	.05	.17	−.12	−.06	.15	−.11	−.15
Corn production	−.35	−.30	−.34	−.33	−.37	.05	.25	−.33	−.25
Number of swine	.00	−.11	.09	.14	.12	−.21	.04	.20	−.04
Number of cattle	.00	.06	.13	−.05	.14	−.45	.26	.28	−.26
Value of livestock	.21	.30	.27	.27	.12	.02	−.20	.19	.20
Value of farms	−.02	−.07	.04	−.26	.18	−.03	.06	−.01	−.06
Value of farms per acre	−.10	−.03	.13	.08	−.25	.12	−.18	.02	.19
Value of farm implements	−.04	−.11	−.08	.03					
Total population	.09	.04	−.14	.02	.32	.19	−.21	.05	.21
Percent slaves of total population	.38	.37	.11	.51	.45	.16	−.43	.30	.43
Percent slaveholders of male adult free population					−.16	−.10	.20	−.06	−.20
	MULTIPLE CORRELATION WITH:								
Multiple correlation coefficient R	.78	.81	.66	.83	.78	.67	.67	.72	.67
Coefficient of determination R²	.59	.65	.43	.68	.61	.44	.45	.52	.45

cate that other factors are involved. For example, the coefficient of determination for the multiple correlation between the economic variables and Whig percentage voting in 1840 is only .59 or 59 percent. Thus about 41 percent of the Whig voting behavior can not be attributed to these economic considerations. Since the rest of the coefficients range from .43 or 43 percent to .68 or 68 percent, one must conclude that other factors equally as important as economic considerations underlay party alignment.

The upper part of the table contains partial correlation coefficients between a single economic variable and voting percentage with all other economic variables controlled for.[12] Thus if there was a connection between political alignment and an economic variable such as percent of slaveholders of the male adult free population, we could expect a high correlation between the two. Inspection of the table discloses that all the coefficients are rather low, with many being near zero. The obvious conclusion is that none of the economic variables was a consistent determinant of voting behavior. Alexander points out that the low correlation coefficients "relating the proportion of slaveholders among the free male adults with political alignment" strongly suggests ". . . that slaveholders as such were no more inclined to be Whigs than Democrats."[13]

What has multiple and partial correlation analysis revealed about political alignment in Alabama before the Civil War? This technique, supplemented by other procedures, demonstrates clearly that the traditional economic interpretation of the composition of the Whig party and the Democratic party in Alabama can not be supported. Alexander takes a positive tack by concluding that the technique does suggest a general conclusion about political alignment and economic development. Although we are not told how he arrived at this conclusion, Alexander argues that Whig party appeals apparently were more effective in those counties with the more advanced stages of economic development. He adds that "It is not that a planter with many slaves or an affluent merchant was more likely to be a Whig than was a carpenter or a small farmer; it is simply that all of these men were more likely to be Whigs if they lived in well developed economic communities rather than in frontier, or isolated, or more nearly self-sufficient communities."[14]

Both bivariate and multiple and partial correlation analysis can be used with profit in historical research. The "pay off" in our example of multiple and partial correlation, as frequently is the case in correlation analysis, is that while it undermined a traditional interpretation it also suggested unexpected patterns and relationships. Once the researcher's attention is directed to these relationships and patterns, he can then begin to ask the right questions. This is simply to say that correlation analysis is not an end in itself.

Guttman Scaling

It will be recalled from the discussion of Guttman scaling in Chapter 4 that the procedure is time-consuming and tedious, especially when a large number of persons and items is involved. However, computers with the capability of performing virtually all the operations of Guttman scaling remove this constraint. Doubtless, this is one of the reasons why historians and political scientists in increasing numbers have utilized the technique in recent years.

One of the earliest published historical studies using the scale analysis technique was William Aydelotte's analysis of "Voting Patterns in the British House of Commons in the 1840s."[15] More recently, two monograph studies utilizing Guttman scaling technique have appeared: Joel Silbey's *The Shrine of Party: Congressional Voting Behavior, 1841–1852*[16] and Thomas B. Alexander's *Sectional Stress and Party Strength: A Study of Roll-Call Voting Patterns in the United States House of Representatives, 1836–1860.*[17] A number of doctoral dissertations in history and political science have been written or are in the process of being written which draw upon Guttman scaling.

One of our illustrations of Guttman scale analysis in historical research is based on Allan Bogue's study of Senate voting in the second session of the 37th Congress.[18] The focus of his investigation was whether or not there were moderate and radical Republican Senators in that session of Congress. To put it more precisely, "Did party or factional groups within the Republican Party the more significantly influence the voting of the Republican Senators" during this session of Congress?[19]

The basic data for the phase of Bogue's study we are describing consisted of 87 roll calls on slavery and confiscation of property legislation. Roll call responses were recorded in punch cards with one card for each Senator. The first 11 columns contained identification information such as, card number, legislators' identification number, Senate identification, session number, and Congress number. Columns 12 to 80 of each card contained the response of each Senator to the roll call votes. A yea vote was coded "1," a nay vote was coded "2," and failure to vote was coded "9." These 87 roll calls were then submitted to a cross tabulation computer program that generated four-fold tables, each of which compared every roll call to every other roll call.[20] The output showed the number and percentage of Senators that voted yea-yea, yea-nay, nay-yea, and nay-nay on the two roll calls under consideration. In addition, the program also calculated Yule's Q coefficient for each pair of roll calls. Each of the 87 roll calls on slavery and confiscation of property was assigned a support or nonsupport position, that is, whether a yea vote or a nay vote meant support or nonsupport. Those roll calls with a 10 percent error or higher in the "error cell" of the four-fold table were rejected. This was not applied when the support position of one roll call was yea and the support position of the second roll call of the pair being compared was nay. In this instance, the error percentage was ignored and the roll call retained.[21] The Q coefficient for each pair of roll calls thus retained was entered into a correlation matrix to indicate similar roll calls. Out of the original 87 roll call votes on slavery and confiscation, a total of 48 met the criterion of scalability and fell into six clusters.[22]

The response pattern of each Senator on each of the six sets of scalable roll calls was then manually posted.[23] This consisted of arranging the roll calls in each set from easiest to support to hardest to support and indicating if a Senator cast a positive vote $(+)$, a negative vote $(-)$, or was absent (0). This information was then used to construct a scalogram for each of the six sets of roll calls.

The one voting pattern reproduced from Bogue's study represents what he calls the dominant voting pattern on all six scales. This pattern scale or scalogram is shown in Table 6.4. The scale type or position numbers are in reverse order from what is normally done, but this is due to the fact that they derive from the first set of roll calls which are arranged in descending order from hardest to oppose to easiest to oppose. The second set contains the same roll calls but with the support pattern. If scale type numbers were assigned to the second set, they would start with 0 at the top and run through 6 at the bottom. It should be emphasized that the first set of roll calls displays the pattern vote of opposition. A $+$ entered for each Senator in this set means that he voted according to the opposition pattern on that roll call. The pattern vote of support for the same roll calls is entered under the roll call numbers in the second set. A $+$ in this set means a Senator voted according to the support pattern on that roll call. A $-$ or x indicates a deviant or "error" vote.

In interpreting this scale, note how Bogue handled the "inconsistent" or "deviant" votes and absences. There are only two "deviant" votes in the scale—the yea vote of Senator Wright on roll call 2 and the nay vote of Senator Wade on roll call 5. In both instances the scale type assignment of each Senator is consistent with his overall pattern. Absences are handled in much the same manner. Since every absence is preceded by a pattern opposition vote, assignments to scale types were made on the reasonable basis that if a legislator registered a pattern vote on a difficult roll call he would do the same thing on an easier one.

Note that there are seven scale types, ranging from strong opponents

TABLE 6.4   A Voting Pattern on Slavery and Confiscation Measures

+ = pattern vote; −, x = deviant vote (error); 0 = absent. Forty-six other roll calls from the original selection fitted this scale. Missing Senators did not vote in half of the divisions shown here. Coefficient of Reproductability = .99

Voting Key: 1—Final vote on S.351, supplementary to the act for the emancipation in the District of Columbia.

2—Final vote on S.394, to amend the act calling forth the militia.

3—Sumner's motion to amend S.351 by inserting, "That in all judicial proceedings . . . there shall be no exclusion of any witness on account of color."

4—Sherman's amendment to S.394, inserting, "who . . . shall owe service or labor to any persons who, . . . levied war or has borne arms against the United States . . . "

5—Sumner's amendment to S.365, providing for emancipation in the state of West Virginia.

6—King's amendment to the confiscation bill, S.151, inserting "persons in the present insurrection levying war against the United States or adhering to their enemies . . . "

SENATOR	PARTY	STATE	TYPE	1 N	2 N	3 N	4 Y	5 N	6 N	1 Y	2 Y	3 Y	4 N	5 Y	6 Y
Powell	D	Ky.	6	+	+	+	+	+	+						
Kennedy	BU	Md.	6	+	+	+	0	+	0						
Davis	BU	Ky.	6	+	+	+	+	0	+						
Wilson, R.	BU	Mo.	6	+	+	0	+	+	+						
Carlile	BU	Va.	6	+	+	+	0	+	+						
Wright	D	Ind.	6	+	−	+	+	+	+			x			
Saulsbury	D	Del.	5		+	0	0	+	+	0					
Stark	D	Ore.	5		+	0	0	+	+	0					
Willey	BU	Va.	5		+	+	+	+	+	0					
Henderson	D	Mo.	4			+	+	+	+	+	+				
Cowan	R	Pa.	4			+	+	0	+	+	+				
Browning	R	Ill.	4			+	+	+	+	+	+				
Anthony	R	R.I.	3				+	+	+	+	+	+			
Doolittle	R	Wis.	3				+	+	+	+	+	+			
Collamer	R	Vt.	3				+	+	0	+	0	+			
Sherman	R	O.	3				+	+	+	+	+	+			
Foster	R	Conn.	3				+	+	+	+	+	+			
Ten Eyck	R	N.J.	3				+	+	+	+	+	+			
Fessenden	R	Me.	3				+	0	+	+	+	+			
Lane, H. S.	R	Ind.	3				+	+	+	0	+	0			
Simmons	R	R.I.	3				+	+	+	+	+	+			
Howe	R	Wis.	3				+	+	+	+	+	+			

From Allan G. Bogue, "Bloc and Party in the United States Senate: 1861–1863," *Civil War History,* XIII (1967), pp. 230 and 232.

TABLE 6.4   (Continued)

SENATOR	PARTY	STATE	TYPE	1 N	2 N	3 N	4 Y	5 N	6 N	1 Y	2 Y	3 Y	4 N	5 Y	6 Y
Harris	R	N.Y.	2					+	0	+	+	+	+		
Foot	R	Vt.	2					+	+	+	+	+	+		
Clark	R	N.H.	1				+			+	+	+	+	+	+
Hale	R	N.H.	1				+			+	+	+	+	0	0
Wilson, H.	R	Mass.	1				+			+	+	+	+	+	+
Sumner	R	Mass.	1				+			+	+	+	+	+	+
Morrill	R	Me.	1				+			+	+	+	+	+	0
Lane, J. H.	R	Kan.	1				+			+	+	+	+	+	+
Harlan	R	Ia.	0							+	+	+	+	0	0
Pomeroy	R	Kan.	0							0	+	0	+	+	+
Grimes	R	Ia.	0							+	0	+	+	+	+
Chandler	R	Mich.	0							+	+	+	+	+	+
Wilkinson	R	Minn.	0							+	+	+	+	+	+
Trumbull	R	Ill.	0							+	0	+	+	+	+
King	R	N.Y.	0							+	+	+	+	+	+
Wade	R	O.	0					x		+	+	+	+	−	+
Wilmot	R	Pa.	0							+	0	+	+	+	0

in type 6 to strong supporters in type 0. Conservative representatives of the border slave states comprise the extreme opposition scale types 6 and 5, while moderate Republicans and war Democrats are in the middle range scale types 4 and 3. Radical Republicans make up scale 2, 1, and 0. The rank order derived from this scale along with a cluster analysis resulted in the classification of radical and moderate Republicans in Table 6.5. Quite obviously Republican Senators were divided,

TABLE 6.5   Republican Radicals and Moderates

REPUBLICAN RADICALS		REPUBLICAN MODERATES	
Chandler (Mich.)	Morrill (Me.)	Anthony (R.I.)	Foster (Conn.)
Clark (N.H.)	Pomeroy (Kan.)	Browning (Ill.)	Harris (N.Y.)
Foot (Vt.)	Sumner (Mass.)	Collamer (Vt.)	Howe (Wis.)
Grimes (Ia.)	Trumbull (Ill.)	Cowan (Pa.)	H. S. Lane (Ind.)
Hale (N.H.)	Wade (O.)	Dixon (Conn.)	Sherman (O.)
Harlan (Ia.)	Wilkinson (Minn.)	Doolittle (Wis.)	Simmons (R.I.)
King (N.Y.)	Wilmot (Pa.)	Fessenden	Ten Eyck (N.J.)
J. H. Lane (Kan.)	Wilson (Mass.)	(Me.)	

From Allan G. Bogue, "Bloc and Party in the United States Senate: 1861–1863," *Civil War History,* XIII (1967), pp. 230 and 232.

but Bogue cautions against concluding that factionalism was of greater consequence than party. He points out that only on 21 of the 87 slavery and confiscation roll calls was there a substantial division between moderate and radical Republicans. Indeed, 65 of these roll calls show a majority of Republicans opposed to a majority of Democrats. Professor Bogue concludes that while much can be said for both the influence of party and factionalism among Republican Senators in the second session of the 37th Congress, his analysis suggests that neither should be emphasized to the exclusion of the other. Our understanding of this Civil War Congress, he says, will benefit more from approaching the political system of this era as one "in which a variety of determinants of voting behavior were interacting with each other. . . ."[24]

A recurring theme in Bogue's study is that historians of legislative behavior should not rely exclusively on Guttman scale analysis in their research. He points out that scale analysis ". . . does not isolate the members of self-conscious groups, and discards roll calls that will not scale."[25] These defects can be corrected in part by using the index number approach and the cluster block technique. For example, had Bogue not used the index number approach that revealed considerable voting along party lines on slavery and confiscation roll calls, he would have concluded from the scale on these roll calls that factionalism was far greater than it actually was.

Our second example of Guttman scale analysis in historical research is taken from Thomas B. Alexander's *Sectional Stress and Party Strength*. The data for this study consisted of the responses of members of the U.S. House of Representatives to selected roll calls of the first session of each Congress that met during the period 1836 to 1860. The analysis was undertaken in order to compare the influence of party and section on congressional voting over an extended period of Democratic Whig rivalry. Guttman scale analysis was utilized because it measures attitudinal responses within a common frame of reference and classifies each respondent in terms of how far he was willing to go in supporting a given policy. Once sets of scalable roll calls were identified and the pattern of scale type voting of each Congressman in each Congress was disclosed, it was a simple matter to arrange the scale types in terms of party and section. Alexander's analysis yielded 33 such scales which related primarily to economic questions and slavery.

The book's appendix contains a listing of the two source programs used for identifying scalable roll calls and the plotting of each individual's response pattern.[26] A writeup accompanies each program so a user should encounter little difficulty in using the programs. Certainly, anyone considering the use of Guttman scale analysis of roll calls for the first time would profit from a very careful reading of this material. A third program is listed which Alexander used to generate sectional pairings on each roll call based on a two-thirds majority vote of each sectional delegation. The output from this program is quite useful in isolating

TABLE 6.6. Economic Scalogram

Scale Position	NE		MA		SNW		NNW		SA		SC		BS		LS		Total	
	D	W	D	W	D	W	D	W	D	W	D	W	D	W	D	W	D	W
0	1		2						2		1[e]		1		2		6	
1	6		12		1				8		1[f]		7		2		28	
2	4		12		1		4		5				1		4		26	
3	6		16		3		1		18		3	1	17	1	4		47	1
4			4		6		1		2	1[d]	4	1	4	1	2	1	17	2
5	1[b]		1					1	1	6	3	7	2	10	2	3	6	14
6		1		5		1		2		4		5		8		1		18
7		13		8		5		1[c]		1						1		28
8		4		8		1				8	2[g]	6[h]	2	13		1	2	27
Total	18	18	47	21	11	7	6	4	36	20	14	20	34	33	16	7	132	90

From Thomas B. Alexander, *Sectional Stress and Party Strength*, Vanderbilt University Press, pp. 10 and 12.

Sectional and Party Distribution of Scale Positions of Representatives, 24th Congress, First Session[a]

[a] Code number and positive position for each roll call in the set (− indicates "nay" = positive position, semicolon indicates scale-position breaks): −46, 23; 50; 40; −45, 61, −28; 21, 20, −52, −33; −47, 44; −16, −43; −22.

Coefficient of reproducibility, .91; four men excluded as nonscale types; ten men excluded because of excessive nonvoting; four men of unknown party affiliation.

[b] N. B. Borden (Mass.)
[c] E. Howell (Ohio)
[d] A. Rencher (N.C.)
[e] R. M. Johnson (Ky.)
[f] C. Johnson (Tenn.)
[g] C. Allan (Ky.), L. Lea (Tenn.)
[h] W. B. Carter (Tenn.), W. J. Graves (Ky.), J. Harlan (Ky.), H. Johnson (La.), J. R. Underwood (Ky.), J. White (Ky.)

patterns of group sectional disagreement on single roll calls and complements the scales described above.

To illustrate Alexander's application of Guttman scale analysis, the first scale listed in the book, labeled an economic scale, is shown in Table 6.6.[27] The procedure followed was basically like that used in Bogue's study. The votes of each Congressman were recorded on separate punch cards using the ICPR code format.[28] These cards were then used as input data for a cross tabulation program which generated four-fold tables, each of which compared every roll call to every other roll call. This program differs from the one Bogue used in that only percentages are listed in each cell of the tables and the Q coefficient is not computed. Then the positive position of each roll call was determined by ascertaining whether a yea or nay vote meant support of the Whig position on economic questions. The 24 roll calls were ranked from the highest percentage support vote to the lowest percentage support vote. The "error" cell in each of the four-fold tables[29] was then inspected and eight roll calls were eliminated because the "error" cell registered more than 10 percent. The remaining 16 roll calls are identified by number in the scalogram and their substantive content as well as the actual vote can be checked by referring to Table 24-5 in the text.

The second computer program was then used to list the responses of each Congressman in terms of his agreement or disagreement with the positive position on each roll call of the scale set. The input data consisted of the original roll call responses used in the first stage of the analysis, plus additional information consisting of the code numbers of the scalable set of roll calls arranged in order from easiest to support to hardest to support. Each roll call number was preceded by a + if the positive position was yea or − if the positive position was nay. The printout from this program listed on the left margin the code number of each Congressman and across the top the code numbers of the roll calls arranged in increasing difficulty to support with a + if the positive position is yea or a − if the positive position is nay. To the right of each member's code number and underneath each roll call is a + if he voted on the positive side, − if he voted on the nonpositive side, and a 0 if he failed to vote.

This information on 240 individual Congressmen was then used to construct the nine scale types shown in Table 6.6. These scale types were then arranged in terms of sectional and party distribution of scale positions. Several general comments about this scalogram are in order. The numbers in the columns represent the number of representatives by party and section who fall into each scale position. The alphabetic subscripts identify those representatives who dissented sharply from either a sectional or a party scale position. Turning now to the question of what the scalogram tells us about the influence of party and section, we note that the extreme scale positions, 0 through 2 and 6 through 8, are occupied principally by Democrats and Whigs respectively. The moderate scale positions, 3 through 5, are occupied by southern Democrats and some southern Whigs. With the exception of the 14 southern Whigs in scale position 8, southern Whigs were more moderate than northern Whigs on these issues. Alexander concludes that "Southerners, whether Democrats or Whigs, were more likely to be drawn from their party position toward the economic middleground, but it is evident that some of the southern representatives were as divided in their economic positions as any in their party."[30]

This division along general party lines on economic questions can be contrasted with the sectional division on roll calls involving slavery questions, as shown in the scalogram in Table 6.7. While a sectional difference among Democrats is evident, it is slight when compared with that of the Whigs. Thus, while Democrats registered considerable unity on slavery questions, the Whigs were sharply divided along sectional lines by slavery issues. As Alexander points out, this sharp sectional division on slavery questions was a "harbinger of Whig party troubles" in the South. [31]

The two examples of the use of Guttman scale analysis in the study of legislative behavior should help introduce the beginner to the technique.[32] However, it should not be assumed that scale analysis must be confined to roll call analysis. Voting patterns on the United States Supreme Court have been analyzed with the scale analysis technique.[33] Anthropologists have found scale analysis useful in measuring the cultural development of societies.[34] A very interesting application of the technique is Peter G. Snow's measurement of political development of Latin American countries.[35] A field yet to be exploited to any extent is that of survey analysis data. This wide range of present and potential applications should suggest ways in which historians can use computer scale analysis in a variety of research applications.

Computer Content Analysis

Verbal data recorded in written documents have been the historian's major source of information about the past. His goal in using these sources has been to draw valid inferences which will help explain and illuminate a given event or series of events in the past. In this sense content analysis of source documents has been a key analytical tool in conventional historical research. While the methodology of this kind

TABLE 6.7.   Slavery Scalogram

Scale Position	NE D	NE W	MA D	MA W	SNW D	SNW W	NNW D	NNW W	SA D	SA W	SC D	SC W	BS D	BS W	LS D	LS W	Total D	Total W
0			4		3				6	6	10	15	7	13	9	8	23	21
1			6		1		1	1[c]	10	6	1	1	8	7	3		19	8
2	15		30		5		4		21	10	2	2	18	11	5	1	77	12
3	1		5	1[b]	3		1	1			1[d]	2[e]	1	2			11	4
4	2	14	1	17		6		1									3	38
5		5		3				1										9
Total	18	19	46	21	12	6	6	4	37	22	14	20	34	33	17	9	133	92

Number of Representatives by Section and Party

From Thomas B. Alexander, *Sectional Stress and Party Strength*, Vanderbilt University Press, pp. 10 and 12.

Sectional and Party Distribution of Scale Positions of Representatives, 24th Congress First Session[a]

[a] Code number and positive position for each roll call in the set (— indicates "nay" = positive position, semicolon indicates scale-position breaks): 53, 38; −26; −37, −10, −15, 1; −43, −36, −9; −35, −11.

Coefficient of reproducibility, .95; two men excluded as nonscale types; nine men excluded because of excessive nonvotings; four men of unknown party affiliation.

[b] D. Wardwell (N.Y.)
[c] D. Spangler (Ohio)
[d] C. Allan (Ky.)
[e] J. Harlan (Ky.), J. R. Underwood (Ky.)

of analysis is seldom explicitly formulated, it seems to rely heavily on intuition and fragmented impressions of a document. Furthermore, controls are seldom used to guard against cumulative impressions of a document which can imperceptibly influence interpretation of that document. In short, content analysis in conventional historical research does not insure systematic reading of a document.

As a corrective to a nonsystematic analysis of verbal data, scholars in communication research tried to develop a systematic content analysis tool. This early effort met with limited success, since the routine clerical operations of a technique confined its application to small data groups. The rapid development of computer technology and the gradual acceptance of computers in social science research have removed this restriction. The coupling of computer speed, accuracy, and repetitiveness of operation with the techniques of systematic content analysis has resulted in two broad categories of computer content analysis. The first type consists of word count programs which generate the frequency with which each word appears in the text. An example of this is Richard Merritt's *Symbols of American Community, 1735–1775* (1966). The second type uses a dictionary containing information about words which reflects the investigator's understanding of the words or those of a panel of experts. The computer compares text words with those in the dictionary and automatically codes the text words with the information in the dictionary. The rationale for a dictionary is that of providing a systematic instrument for reducing the hundreds and thousands of different words in a text to a smaller and more manageable number of categories of meaning. For example, if one is interested in an economic theme, he can find certain entry words in the dictionary which reflect this dimension. Every time a text word is read which corresponds to one of these dictionary words, it is subsumed under the economic theme. This is a tremendous improvement over mere frequency counts of text words.

Probably the most widely used computer content analysis embodying the dictionary concept is that known as the General Inquirer System.[36] This system, which has been revised several times, was developed at the Harvard Laboratory for Social Relations by Philip J. Stone, Dexter C. Dunphy, Marshall S. Smith, Daniel M. Ogilvie, and their associates. While the General Inquirer System has several dictionaries, the one known as the Harvard III Psycho-sociological Dictionary may be of greatest immediate use to historians.[37] This particular dictionary draws heavily on the role theory of sociology and psychological themes of social-psychology. Since some areas of historical research employ concepts, if not assumptions, from sociology and psychology, it seems sensible to use a dictionary which derives its theoretical underpinnings from these disciplines.

The Harvard III Dictionary contains about 3600 entry words which are classified in terms of 83 categories. These 83 categories are divided into two sets of tag groups—first order and second order. Each entry word is assigned a single first order tag which represents the "primary explicit denotative" meaning of that word. Table 6.8 discloses the 55 first order tag categories under 11 major headings: persons, groups, physical objects, physical qualifiers, environments, culture, emotion, thought, evaluation, social-emotional actions, and impersonal actions. The re-

TABLE 6.8 Harvard Third Psychosociological Dictionary: List of Tags

FIRST ORDER TAGS

Persons		*Emotions*	
self	01	arousal	27
selves	02	urge	28
other	03	affection	29
male-role	04	pleasure	30
female-role	05	distress	31
neuter-role	06	anger	32
job-role	07		
Groups		*Thought*	
small-group	08	sense	33
large-group	09	think	34
		if	35
Physical objects		equal	36
body-part	10	not	37
food	11	cause	38
clothing	12	def-mech	39
tool	13		
natural-obj	14	*Evaluation*	
non-spc-obj	15	good	40
		bad	41
Physical qualifiers		ought	42
sensory-ref	16		
time-ref	17	*Social-emotional actions*	
space-ref	18	communicate	43
quan-ref	19	approach	44
		guide	45
Environments		control	46
social-plac	20	follow	47
natur-world	21	attack	48
		avoid	49
Culture			
ideal-value	22	*Impersonal actions*	
deviation	23	attempt	50
action-norm	24	get	51
message-form	25	possess	52
thought-form	26	expel	53
		work	54
		move	55

TABLE 6.8. *(Continued)*

SECOND ORDER TAGS

Institutional		*Psychological*	
academic	56	overstate	71
artistic	57	under-state	72
community	58	sign-strong	73
economic	59	sign-weak	74
family	60	sign-accept	75
legal	61	sign-reject	76
medical	62	male-theme	77
military	63	female-theme	78
political	64	sex-theme	79
recreational	65	ascend theme	80
religious	66	authority-theme	81
technological	67	danger theme	82
		death-theme	83
Status			
higher-stat	68		
peer-status	69		
lower-status	70		

SAMPLE EDITED TEXT

Declar01	We hold these truths to be self-evident, that all	(card 1)
—	men are created equal, that they are endowed by their	(card 2)
—	Creator with certain unalienable Rights, that among	(card 3)
—	these are Life, Liberty, and the Pursuit of Happiness.	(card 4)
—	That to secure these rights, Governments are instituted	(card 5)
—	among men, deriving their just powers from the consent of	(card 6)
—	governed + that whenever any form of government becomes des-	(card 7)
—	tructive of these ends, it is the right of the people to alter or abolish it,	(card 8)

maining 28 tags are assigned to a second order classification which represents the connotative meaning of dictionary entries. These second order tags are subsumed under three major headings: institutional context, status context, and psychological themes. A dictionary entry may be assigned as many second order tags as are necessary to give a satisfactory definition. Note that Table 6.8 also displays the 26 higher order tags and some sample edited text. It might be noted in passing that users of this dictionary report that it tagged between 95 and 98 percent of the text.

Textual material that is to be submitted to computer content analysis must be transcribed into punch cards. The General Inquirer System permits a wide range of text editing.[38] The minimum editing required consists of assigning an identification code to each document and using certain keypunch conventions. Editing may also involve syntactic identification, which uses four syntax identifiers—subject, verb, object, and unclassified. When a syntax subscript follows a text word, a syntax marker (S, V, O, U) is added to the tag category which computer operations automatically assigned to that word. Another form of editing is that of adding descripters to the text to define pronouns or explain words. Table 6.8 shows a sample of edited text. Each line of text is recorded in columns 12 to 80 of an IBM card. The first ten columns are reserved for an identification code. In the example the identification code consists of an abbreviation of the author's name and a key to where that section of text is located in the original document. The hyphen in the first column of those lines immediately under the identification code instructs the computer to assign this code to these lines of the text. This is a convenience for keypunch operators and in no way affects analysis of the document.

As suggested earlier, the initial phase of computer content analysis involves comparing each text word with dictionary entries. When there is a match, the text word is assigned the tag number of the dictionary entry. The output from this program is written on magnetic tape and serves as input for a tally program which generates a frequency count of each tag and calculates an index score for each tag. This index score can represent the number of tag assignments as a percentage of the total number of words in the document or the number of sentences in the document. This same output can be used to retrieve sentences which contain tags specified by the investigator. For example, one could retrieve all sentences which contain both a time reference tag and an ideal-value tag. The level and complexity of the retrieval questions are determined by the investigator.[39]

With this rather brief description of the General Inquirer System behind us, attention can now be turned to an illustration of its use in historical research. The example is based on a paper read by James Henderson and Charles Dollar at the Wisconsin Political and Social History Conference.[40] The point of departure of this paper, which was primarily a report on research in progress, was that of the social and psychological stress of the rhetoric of the American Revolution, an emphasis which Gordon Wood recently suggested ought to be explored.[41] Hence, attention was directed to 30 of the 83 tags of the Harvard III dictionary which might measure such a dimension.

Three documents written between 1767 and 1776 were selected in order to measure this social and psychological stress. The documents are John Dickinson's *Letters from a Pennsylvania Farmer,* written in 1767, Thomas Paine's *Common Sense,* and the *Declaration of Independence,* both written in 1776.[42] The three documents have a number of things in common: each represented the patriot position; each was issued at a critical moment in the movement toward independence; each was widely circulated and came to represent an official or quasi-official patriot position. On the other hand, the men who wrote the documents are in many ways dissimilar. Dickinson was a moderate who opposed the decision to declare independence; Paine and Jefferson were both radicals in the sense that they both advocated separation from the mother country yet they were different in their backgrounds. Paine was an English immigrant to Pennsylvania, attracted by the prospect of Revolution and Jefferson was a member of the Virginia aristocracy. The documents they wrote are also conventionally judged dissimilar. Dickinson's *Letters from a Pennsylvania Farmer* were written at the time of the Townshend Acts crisis in 1767, and are generally construed to represent the constitutional position of the colonies at that time, while Paine's *Common Sense* is best known as the pamphlet which delivered an incendiary attack against George III, the institution of monarchy, and the colonial connection with England. The *Declaration of Independence* is understood variously as a statement of Enlightenment philosophy justifying revolution or a legal brief for colonial independence.

Henderson and Dollar emphasize that an analysis of the social and psychological stress of the rhetoric of the Revolution suggests the use of the Harvard III dictionary. Most historians, they say, are not alert to some of the themes tagged in the dictionary such as "natural world," "deviation," "distress," "danger," and "death"—at least not in the systematic fashion attempted in the General Inquirer approach where analysis focuses on both denotation and connotation of the single word. This dictionary should, ideally, allow for a more sensitive analysis of Revolutionary literature, both of general tendencies over time and of comparisons of varieties of Revolutionary rhetoric as articulated by moderates and extremists as the terms have been conventionally understood.

Within this context, the output of a computer content analysis of the three documents listed in Table 6.9 is quite interesting. The 30 tags listed here related to what one might call the mood of the documents. Despite the differences among the documents and authors mentioned earlier, some of the index scores for the 30 tags when plotted on a graph are similar in direction. For example, on the political tag all three documents have a high score (9.3 to 10.3), while on the attempt tag all three have extremely low scores. Since all three documents deal with political questions, the high scores on this tag are not unexpected. More interesting, however, are the index scores for those tags which seem to reflect social and psychological stress emanating from a doctrinaire position. The mean index scores of this dimension for the two documents generally classified as propagandistic, Dickinson's *Farmer's Letters* and Paine's *Common Sense,* are 1.53 and 1.81 respectively. This is not unexpected, since Paine is recognized as an ideologue. However, when this mean index score of *Common Sense* is compared with that of the *Declaration of Independence*, the latter shows a much higher score, 2.94. How is it possible that the *Declaration of Independence,* "a legalistic" bill of indictment against the king "written in the language of British constitutionalism," to use Daniel Boorstin's words, could surpass in polemic force the rhetoric of Paine, the full blown ideologue? A retrieval of the selected tags in the *Declaration of Independence* discloses that the language of the document is in fact far from "legalistic" or technical. The assertion that the Quebec Act was a policy "for abolishing the free system of English laws in a neighboring Province, establishing therein Arbitrary government, and enlarging its boundaries so as to render it at once an example and fit instrument for introducing the same Absolute rule into these colonies," for example, is scarcely a complaint couched in constitutional technicalities. Furthermore, such language as "He has plundered our seas, ravaged our coasts, burnt our towns, and destroyed the lives of our people" is more than a mere point of law. In short the *Declaration of Independence* is couched in a rhetoric of impassioned radicalism which is frequently overlooked. This suggests that Gordon Wood and Bernard Bailyn are correct in stressing the ideas of the American Revolution as reflected in its rhetoric. In addition, the results of this analysis indicate that Wood's suggestion that the Revolution be examined in the context of a generalized revolutionary syndrome has considerable merit and should be undertaken.[43]

The authors concluded their paper with several observations about the advantages of computerized content analysis in general and the Harvard III Dictionary in particular which historians contemplating its use may find helpful. A basic point repeated several times is that computerized content analysis can not substitute for sensitive reading of documents—it can only enhance that reading. By focusing on single words in a perfectly consistent manner, computerized content analysis can help avoid cumulative intuitive impressions of a document and thereby direct attention to themes which may have been overlooked or were partially obscured by initial impressions.

Parallel to this is the conceptualization of the tags in the Harvard III dictionary. The organization of tags along dimensions of meaning which reflect social-psychology theory can be quite informative to historians engaged in research involving intergroup relationships and their perception of issues.

A note of warning is directed to historians who may be tempted to accept uncritically the results of computer content analysis.[44] Mistagging can occur since the context of the word may alter its meaning, and the computer will miss this subtlety. For this reason extensive re-

TABLE 6.9   Selected Tag Index Scores of *Letters from a Pennsylvania Farmer, Common Sense,* and the *Declaration of Independence*

TAGS[a]	DOCUMENTS		
	DICKINSON	PAINE	JEFFERSON
First Order Tags			
Natur-World	.7[b]	1.6	1.2
Ideal-Value	3.6	4.3	3.9
Deviation	.3	1.0	.9
Action-Norm	1.8	1.2	1.8
Distress[c]	.6	.8	.7
If	1.5	1.2	.6
Not	1.5	2.2	.4
Good	.9	1.1	1.5
Bad	.3	.4	.2
Control[c]	1.1	.8	1.6
Attack[c]	.9	1.0	2.5
Attempt	.5	.5	.5
Expel[c]	.2	.3	.9
Second Order Tags			
Community	3.1	4.3	6.7
Economic	3.4	2.4	1.1
Legal	2.0	1.5	3.2
Military[c]	.9	1.1	2.5
Political	10.2	9.1	10.3
Overstate[c]	3.7	4.3	4.6
Under-state	2.1	2.5	.8
Sign-strong[c]	4.1	4.8	8.0
Sign-weak	1.1	1.1	1.7
Sign-accept	2.3	2.5	2.2
Sign-reject[c]	2.5	3.0	5.5
Authority-theme	7.4	7.2	8.8
Danger-theme[c]	.9	.8	1.6
Death-theme[c]	.4	1.2	1.5
Religious	.6	.9	.5

[a] A brief description of these tags is given in *The General Inquirer*, pp. 174–176.

[b] Tag assignments as a percentage of the total number of words in each document.

[c] Selected tags relating to social and psychological stress.

trieval should be made to check the accuracy of tag assignments. More important, however, is that the Harvard III dictionary is designed for twentieth century literature, not eighteenth century literature. Accordingly, effective use of the Harvard III dictionary requires additional entries into the dictionary as well as altering the tags of words that clearly are not valid for the eighteenth century. In order to make these additions and corrections, it is necessary for users to immerse themselves in the dictionary. In addition, it is highly advisable, the authors say, for historians contemplating use of the Harvard III dictionary, or any other content analysis dictionary, to tag manually textual material. This will give one a feel for the dictionary and an appreciation of the rationale of tag assignments to words.

Since much of the historical data available to historians is in textual and verbal form, it would appear that computerized content analysis can be of great assistance in historical research. This is particularly true of the Inquirer II System which allows for more elaborate searches of the data than the General Inquirer.[45] Inquirer II, which resulted from the collaboration of Dennis J. Arp, J. Philip Miller, and George Psathas is written in PL/1 for the IBM 360 series computers (or a computer with a PL/1 compiler and sufficient core storage). The system is far more flexible than the General Inquirer and permits a wide range of options. Furthermore, Inquirer II incorporates much of the experience derived from solving problems encountered in using the General Inquirer. Historians considering the use of computerized content analysis might begin with the conceptual framework of the General Inquirer. However, the set of computer programs used should be those of Inquirer II, given access to adequate computer capability.

Cluster Analysis

The most difficult computer classification technique is cluster analysis. In Chapter 4, cluster analysis was discussed briefly as a technique for identifying legislators who constituted voting blocs. Cluster analysis techniques can be used in areas other than roll call analysis. Indeed, they are appropriate where the goal is to identify "real types" and describe them in terms of similarity and dissimilarity. "Real types" as opposed to "ideal types" attempt to mirror reality. When one speaks of identifying and describing "real types," the emphasis is directed toward attributes which can be observationally characterized as a profile. Accordingly, a profile score for each subject consists of a number of scores on a certain number of variables. Generally speaking, the construction of "real types" involves answering the following profile questions: (1) How much does the profile of one subject in a data set resemble any other in the same data set? (2) What and how many are the groupings of profiles with high mutual resemblance? (3) How can one express with some precision the central profile tendency of each group, especially when the data set is large?[46]

Cluster analysis techniques are appropriate in answering these questions in historical research. We emphasize this because factor analysis techniques are inappropriate for classifying "real types." Raymond B. Cattell, who is probably the leading American exponent of factor analysis techniques in experimental psychology, notes that factor analysis and cluster analysis are two distinct techniques which accomplish different ends.[47] Even though Cattell directed this comment to researchers in psychology, it is relevant to historians.

There are several differences between cluster analysis and factor analysis which must be kept clearly in mind. First of all, the two techniques make different assumptions about data.[48] In cluster analysis data are treated as "surface traits" which literally describe observed behavior only. Factor analysis starts with the same "surface trait" data, but with this difference—"surface traits" are assumed to be generated by latent "source traits" which are not directly observable.

Another important distinction between the two techniques, which is an extension of the previous point, is the question of what they reveal. Cluster analysis indicates only that certain kinds of observed behavior or attributes go together. To put it another way, cluster analysis discloses only the extent to which certain specified patterns are present or absent among given subjects. In contrast, factor analysis goes beyond attribute relations to disclose mathematical relations called factors which are construed as hypothetical entities that underlie the observed behavior. The factors or hypothetical constructs thus revealed lie beyond immediate observation and yet are considered as real as that which is immediately observed. Furthermore, the meaning of factors is not self-evident, that is, the investigator must identify them and assign appropriate descriptive tags. This is usually done by examining the observed characteristics for patterns. Such factors or constructs in fact approximate what Max Weber called "ideal types" and are not a description of reality as such.[49]

A third distinction between cluster analysis techniques and factor analysis techniques is causation. Cluster analysis indicates only the presence or absence of a pattern among subjects. It does not yield any knowledge as to why such a pattern appeared or whether it is due to a single common influence or to many influences superimposed.[50] On the other hand, the initial assumption factor analysis makes about data implies a causal connection between the mathematical entities called factors and the observed temporal regularity of phenomena.[51] While it is not our intent to get into the thorny philosophical problems of causation in the explanation of historical events, it is sufficient for our purposes to note that most if not all historical explanations either implicitly or explicitly draw upon causal connections among phenomena.[52] However, historians will have a much sounder empirical justification for causal connections, either explicit or implicit, by emphasizing empirically discernible features of a subject matter rather than by inferring from the constructs of factor analysis.

In addition to the above differences between cluster analysis and factor analysis we may distinguish between the two techniques in terms of measures of association. Factor analysis begins with a matrix of correlation coefficients which measure the degree of similarity and dissimilarity among the variables listed across the rows and columns of the matrix. Generally, the Pearsonian product moment correlation coefficient is used in factor analysis, though any coefficient appropriate for the data may be used. The elaborate mathematical operations of factor analysis which extract factors from the matrix require the kind of measure of association that only correlation coefficients yield.[53] Cluster analysis, however, can use a matrix of correlation coefficients, percentage agreement scores, or actual numerical agreement scores say among legislators on roll call votes. This capacity permits a researcher some flexibility in choosing the measure of association appropriate to his data and research design.

After this digression into the appropriateness of cluster analysis techniques vìs-a-vìs factor analysis techniques, let us turn to two illustrations of computerized cluster bloc analysis techniques.

The first example illustrates the use of the Rice cluster bloc analysis technique in studying legislative behavior.[54] The Rice technique, which has been modified considerably since its introduction in 1927, consists of computing agreement scores between pairs of legislators and grouping together those pairs which share a minimum level of agreement.[55] The agreement score can be the actual total number of identical votes cast, a percentage of identical votes out of the total votes cast, or computations which take into account absences, paired votes, and the like. After agreement scores have been calculated for each pair of legislators, they are entered into a matrix in which each row and column represents a single legislator. The highest agreement score between a pair of legislators is entered into the top left corner of the matrix, followed by the next highest score that includes one of the legislators previously listed. This procedure is followed until the legislator having the lowest agreement score with one of the two persons in the initial pair occupies the last column on the right and the last row on the left. Clusters of legislators can then be grouped together in terms of those pairs which share a specified level of agreement.

The matrix of paired agreements of southern Senators on farm bloc roll calls in the 67th and 68th Congresses displayed in Table 6.10 illustrates this procedure. The highest percentage of agreement between any pair of Senators was registered by Simmons and Overman. The names of the other southern Senators are entered across the top and bottom of the matrix in descending order of agreement with Simmons. The blanks in the matrix indicate those pairs of Senators with less than 76 percent of agreement. If one defines bloc membership so as to include only those pairs of Senators who agree with each other on at least 80 percent of the roll calls, then four blocs can be identified. Bloc 1 is made up of Simmons, Overman, Smith, McKellar, Harris, Trammell, and Caraway. Robinson, Harrison, and Heflin form bloc 2. Blocs 3 and 4 are comprised of Swanson and Glass, and Shields and Underwood, respectively. Sheppard and Fletcher can be considered marginal members of blocs 1 and 2 respectively. Even though Broussard, Ransdell, and Dial fail to qualify as members of any bloc, their pattern of voting, along with that of Underwood and Shields, indicates fragmented and limited opposition to blocs 1 and 2, and to a lesser extent, bloc 3.

James Henderson used the Rice cluster bloc analysis technique in his excellent study of partisan conflict in the Continental Congress, 1778–1779.[56] He notes that generally speaking historians have accounted for the emergence of the two party system in the late 1790s as an extension of the dimensions of factional groupings and issues which produced partisan conflict in the Continental Congress. Indeed, the nationalist historians of the nineteenth century tended to relate the factional struggles of the Revolutionary era to the controversies of the Constitutional and Federalist periods. According to this viewpoint, Congressional politics was a contest between the advocates and opponents of central authority, and thus was a prelude to the trials of American federalism. In contrast, twentieth century historians like Carl Becker and Merrill Jensen argue that the socioeconomic underpinnings of Revolutionary politics resulted in the supporters of central authority being primarily concerned with preserving the status quo and its opponents being primarily concerned with fulfilling the democratic promises of the Revolution. Thus the ratification of the Constitution represented the consummation of the conservative reaction against the democratic aspirations of the Revolution. Henderson correctly points out that these two schools of thought employ a priori assumptions about the composition of partisan groupings and the issues which contributed to their formation. Therefore, his computer-assisted cluster bloc analysis of the full voting record of the Continental Congress for a two-year period was designed to uncover the partisan structure of national politics.

The years 1778 and 1779 were selected for analysis for several reasons. Roll call votes were not recorded in the *Journal* until late 1777. There were more roll call votes in 1779 than in any other year. Finally, partisan feelings about both foreign and domestic policy were especially intense in these two years. During these two years 444 roll call votes

TABLE 6.10 Matrix of Paired Agreements among Southern Senators on All Farm Bloc Roll Calls in the 67th and 68th Congresses

Senators Watson, George, Williams, Stephens, Culberson, and Mayfield are not included in this matrix because they were not in office in both the 67th and 68th Congresses. The blanks indicate agreement of less than 76 percent.

	Simmons	Overman	Smith	McKellar	Harris	Trammell	Robinson	Caraway	Harrison	Heflin	Fletcher	Swanson	Sheppard	Glass	Dial	Ransdell	Underwood	Broussard	Shields
Simmons	0	97	96	92	91	90	90	89	88	87	83	80	78	75	73	65	64	61	56
Overman	97	0	94	92	87	85	83	85	83	83	79	78	83						
Smith	96	94	0	84	88	80	86	93	83	83	79	76	78						
McKellar	92	92	84	0	88	82	78	86	78	78	77	77	78						
Harris	91	87	88	88	0	95	85	88	90	91	86	85	81						
Trammell	90	85	80	82	95	0	83	85	88	90	83	84	79						
Robinson	90	83	86	78	85	83	0	90	89	86	86	78	79			78	77		
Caraway	89	85	93	86	88	85	90	0	85	87	84	83	78	79					
Harrison	88	83	83	78	90	88	89	85	0	89	83					78			
Heflin	87	87	83	78	91	90	86	87	89	0	77	79							
Fletcher	83	78	79	77	86	83	86	84	83	77	0								
Swanson	80	78	76	77	85	84	78	83		79		0		93					
Sheppard	78	83	78	78	81	79	79	78					0						
Glass	75							79				93		0					
Dial	73														0				
Ransdell	65						78		78							0		79	
Underwood	64						77										0		86
Broussard	61															79		0	
Shields	56																86		0

were recorded in the *Journal,* and of these 355 were actually used in the analysis. The 89 roll call votes not used were rejected because they involved 90 percent agreement or more, or could be considered duplicate votes since they occurred on the same questions with little or no change in voting.

The voting records of 47 delegates in each year were recorded on punch cards, with each card representing an issue and the vote of each delegate on that issue occupying a column in the card. (Recall that Alexander and Bogue used separate cards for each legislator, not for each roll call; either approach can be used.) The number of delegates was restricted to 47 because this was the maximum number the computer program could handle. This was not a serious problem since it excluded only those delegates who voted on less than 25 percent of the roll calls. The set of cards for each of the two years was used as input data for a computer program which cross tabulated each delegate's yea and nay votes with the yea and nay votes of the other 46 delegates and yielded a percentage of agreement and disagreement for each pair of delegates. These percentage agreement scores were then entered into a 47×47 matrix similar to the one shown in Table 6.10. The matrix of agreement scores for each of the two years was used to identify the bloc structure of the Continental Congress for 1778 and 1779 respectively. These two bloc structures are displayed in Table 6.11 and 6.12.

Before looking at these two tables in detail, it is necessary to give some attention to Henderson's definitions of a voting bloc, "fringe" bloc membership, and "coalitions." A voting bloc is a cluster of delegates who agreed with each other on at least 75 percent of the votes they cast in common. Since some delegates registered a 75 percent level of agreement with some members of a bloc and not with others, a category of "fringe" membership was established. A delegate who agreed 75 percent of the time with at least one half the members of a bloc was designated a fringe member. A coalition is a combination of blocs with overlapping membership. In addition, a coalition lacks interrelated blocs with other coalitions. This is clearly shown in Table 6.11 which indicates a tendency toward development of two opposing alliances or coalitions.

Inspection of Tables 6.11 and 6.12 discloses several interesting dimensions of the bloc structure of the Continental Congress. For one thing, the composition of blocs during the two years varied. This is easily accounted for by the fact that some delegates attended for only one year while others attended for both years but qualified for membership in a bloc for only one year. Perhaps the most striking feature about the two tables is that sectionalism was the most decisive determinant of voting patterns for both years. During the two years no New England delegate was in any bloc dominated by southerners, and only in 1779 was there a southerner (Richard Henry Lee) in a bloc of New Englanders. This sectionalism was intensified in 1779 when bloc delegates from the middle states coalesced into a loose coalition. Despite this, the middle states delegations still inclined toward alignment with the other two sectional coalitions with bloc 3 leaning toward the New England coalition and bloc 4 leaning toward the southern coalition. If coalition cohesion is defined in terms of overlapping bloc membership, then New Englanders were much more cohesive than were southerners. For example, in 1778 nine of the fourteen New England delegates included in the analysis were either core or fringe members of the New

England coalition, and five of these belonged to blocs 1 and 2. In contrast, only 11 of the 21 southerners included in the analysis were either core or fringe members of the southern coalition, and only three of these belonged to blocs 5 and 6. This tendency was more pronounced in 1779.

Henderson concluded that these bloc formations raised at least two basic questions: Why was the New England coalition more cohesive than delegations from other sections? and what caused the middle states to divide as they did? In answering the first question Henderson points out that while economic interest undoubtedly played a major role in cementing the New England alliance, economic interests were not as great as they were in 1779, yet the coalition was about equally cohesive in both years. In addition, economic motives fail to account for the presence of Lee, McLene, Searle, and Shippen (all from Pennsylvania) in the coalition of 1779. A comparison of the involvement of delegates in supporting the Revolution from 1776 indicates that the New Englanders clearly had the strongest connection with the early movement for independence. Furthermore, the delegates in the New England coalition were on the average the oldest group in the Congress. The average age of the New England coalition was 50 years in 1778, while the members of the southern coalition averaged just 39 years in 1778. Also, most of the delegates from the middle states who joined the New England coalition were older than the average.

The cohesiveness of the New England delegation and the question of factional struggles in Congress cannot be reduced to a struggle between young progressives and old revolutionaries. Yet the fact that one group of voting blocs was composed of older men with records of involvement in the early formulations of national policy, while the other group of blocs was composed largely of younger men with limited tenure in the move for independence, can not be ignored.

Henderson attempted to get at the second question raised by the bloc formation, the tendency of the middle states to divide as they did, by looking at those votes on which the middle states were most seriously divided. Sixty-five votes cast during 1779, the year in which the middle states split very sharply, were selected. Twenty-four of these votes (or 37 percent) were concerned with foreign policy. While several issues were caught up in foreign policy the most crucial one was the Lee-Deane imbroglio. The voting alignment on this issue very closely approximates the bloc structure derived from analysis of the full voting record for 1778 and 1779. Henderson points out that while the Lee-Deane affair did not create factions in Congress, it is likely that it reflected already existing tensions. Included among those tensions were the allegations of the New Englanders that Silas Deane and his supporters were using the Revolution for personal advantage and the fears of New Eng-

TABLE 6.11 Bloc Structure of the Continental Congress in 1778 (129 Roll Calls)

NEW ENGLAND COALITION (BLOCS 1–3)

State	1	2	3
N.H.	Bartlett*, Whipple	Whipple*	
Mass.	S. Adams, Holten, Lovell	S. Adams*	
R.I.	Ellery*		Ellery*
Conn.	Dyer, Ellsworth		Dyer, Ellsworth, Sherman*
N.J.	Scudder, Witherspoon		Scudder, Witherspoon
Pa.			Clingan, Roberdeau

SOUTHERN COALITION (BLOCS 4–6)

State	4	5	6
N.Y.		G. Morris*	Duer*
Pa.	R. Morris*	R. Morris	
Del.	McKean		
Md.	Carroll, Chase*, Henry*, Plater*	Carroll, Henry, Plater	Carroll, Plater*
Va.	F. L. Lee*, T. Adams	T. Adams	T. Adams
N.C.		Harnett*	M. Smith, Harnett, Burke
S.C.			Mathews
Ga.			Langworthy

From James Henderson, "Partisan Politics in the Continental Congress, 1778–1779: A Quantitative Approach," an unpublished paper.

Notes: Blocs 1–6 represent clusters of interrelated pairs of delegates who agreed with each other at least 75 percent of the time unless marked by an asterisk (*) which represents a "fringe" member of the bloc who agreed with at least half the "core" members 75 percent of the time.

TABLE 6.12 Bloc Structure of the Continental Congress in 1779 (226 Roll Calls)

From James Henderson, "Partisan Politics in the Continental Congress: 1778–1779: A Quantitative Approach," an unpublished paper.

	NEW ENGLAND COALITION (BLOCS 1–2)		MIDDLE STATES BLOCS (3–4)		SOUTHERN COALITION (5–8)			
	1	2	3	4	5	6	7	8
N.H.	Whipple, Peabody*							
Mass.	S. Adams, Lovell	S. Adams*						
R.I.	Marchant	Marchant						
Conn.	Spencer*	Spencer, Huntington, Sherman						
N.Y.				Duane, Lewis, Fell, Floyd		Jay, Morris		
N.J.			Houston, Scudder, Witherspoon					
Pa.	McLene, Searle, Shippen		Searle*, Shippen	Atlee*				
Del.				Dickinson*				
Md.					Paca, Carmichael, Plater*, Jenifer			
Va.	R. H. Lee					M. Smith*	M. Smith, Griffin*	M. Smith*
N.C.						Penn	Penn, Burke*, Harnett, Hill	Penn, Burke*, Harnett, Hill, Sharpe
S.C.								Drayton*

* Fringe members. (See notes to Table 6.11.)

landers that the French alliance would entangle the American Revolution with old world corruption.

In concluding his study Henderson suggests that the dimensions of partisan politics revealed by cluster analysis fail to support either the class conflict or the centralism–states rights dichotomy of the traditional interpretations of national politics during the American Revolution. An interpretation of national politics which conforms to the sectional dimensions of partisan politics, he says, will focus more on social and cultural tensions before the Revolution and less on expectation of the political and economic aftermath of the Revolution.

An improvement on cluster bloc analysis is what Louis I. McQuitty calls elementary linkage analysis.[57] One reason for this is its definition of a type or cluster. A type is defined by elementary linkage analysis as a group of variables of such a nature that each one in it is more like at least one other variable in this group than any other variable in any other group. In actual practice it usually turns out that if members of each type share sufficient similarity among themselves that there is little overlap between types. The second useful feature of elementary factor analysis is that it yields the identification of a prototype variable that *best* represents the characteristics the members of a given type have in common. Furthermore, when there is considerable similarity among several types this technique permits their combination into higher order types.

Like cluster bloc analysis, elementary linkage analysis requires that measures of association among variables be computed. The coefficients are entered into a matrix and the highest entry in each column is noted. Then the highest measure in the entire matrix is identified and the two variables it represents form the basis for identifying members of the first type. This is done by noting if the highest entry in any column occurs on the row of either of the two variables. If so, these variables are added to the type as first order members. This same procedure is repeated for the rows of the first order type members, and if necessary, for second, third, and fourth order type members, until all the possible additions to the type have been identified. With the identity of this type now established, the same procedure is repeated for the next highest coefficient in the entire matrix which has not been used. This process is repeated until every variable is assigned to a type.

Once the members of each type have been identified, the selection of a representative or prototype variable may be undertaken. The identification of the variable that *best* represents the central tendency of type members is obtained by taking the correlation matrix of each type and summing the column entries for each variable. The prototype variable is the one with the highest column sum. The remaining variables in each type can then be ranked in terms of their correlation coefficient with the prototype.

These procedures become very tedious and time-consuming when the matrix is larger, say, than 30 by 30. One way to avoid this is to use a computer program that prints out on each page beginning with variable 1 the correlation coefficients in sequential order and rank order. Thus on page 1 of the printout are listed the coefficients between variable 1 and 2, 1 and 3, and so forth along with a rank ordering list of the correlation coefficients between variable 1 and all other variables. Each succeeding page of printout lists the same information for each variable until the computations for all variables are listed. It is a simple matter then to locate in the rank order listing on each page of printout the highest coefficient between each pair of variables.

It is helpful to write down on a separate sheet of paper the highest coefficient between each pair of variables, being careful to note the variable numbers. This information can then be schematically arranged by types as in Figure 6.1. A matrix for each type can then be easily con-

Figure 6.1.   Schematic arrangement of hypothetical types

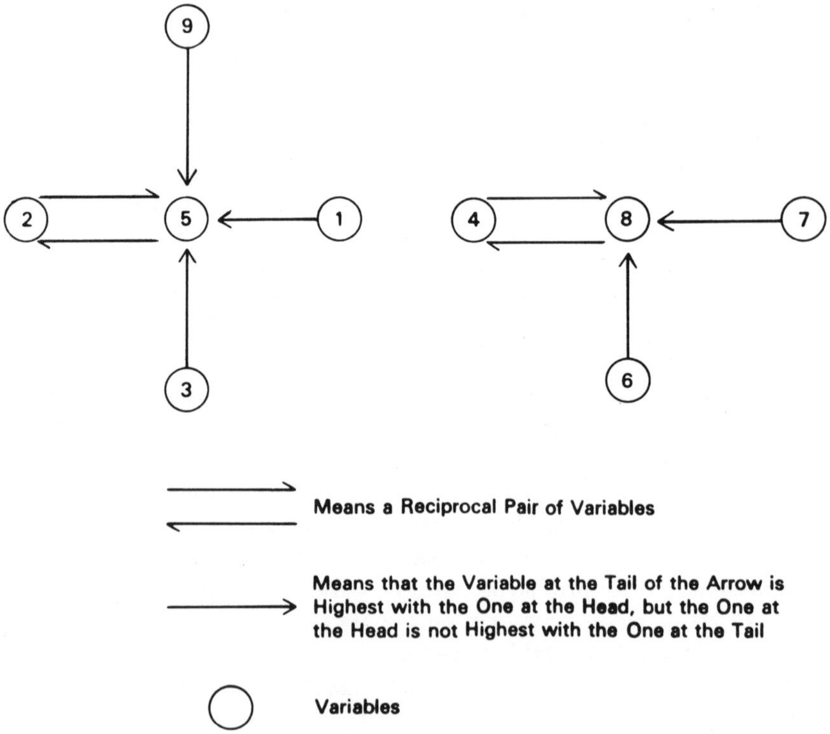

_____→
←_____   Means a Reciprocal Pair of Variables

_____→   Means that the Variable at the Tail of the Arrow is Highest with the One at the Head, but the One at the Head is not Highest with the One at the Tail

◯   Variables

structed from the original computer printout and the prototype variables identified. The time required to do this paper and pencil work depends upon the number of variables handled. Experience indicates that 96 variables can be classified into types and the prototype variable for each type identified in less than five hours. Of course this entire procedure could be programmed, but the paper and pencil work has the advantage of keeping the researcher close to his data.

After this brief introduction to elementary linkage analysis, let us consider its use in resolving a question in political history, that of identifying a critical election in the state of Oklahoma.[58] According to the late V. O. Key, Jr., a critical election is one in which there is a realignment of the elements making up the electorate that issues in new electoral groups and loyalties that persist over time.[59] Key developed a model of critical elections based on Massachusetts as a case study and suggested that 1928 was such an election. In terms of Key's model if 1928 were an election that decisively marked a reshuffling of the electorate's interests and loyalties, it should show up in a redistribution of the presidential vote for that year which persisted through several elections. The capacity of elementary linkage analysis for clustering variables (elections) which share certain attributes (varying Democratic percentage of the vote) makes it particularly useful in isolating a critical election and identifying trends. For if a realignment did occur in 1928, then those elections before this date should cluster together and those elections after this date should likewise cluster together.

The specific question in our example is whether or not the general voting trends in presidential elections, and especially the 1928 election in Oklahoma, conform to Key's model. The data are the Democratic percentage of the two party vote cast in each of the 77 Oklahoma counties in each presidential election from 1908 to 1960, recorded on punch cards with one card for each county. Each election occupied the same columns in all cards. For example, the Democratic percentage for each county in the 1908 election filled columns 11–14 of all 77 cards. A computer calculated a correlation coefficient for every possible combination of pairs of elections, a total of 91, and arranged them in a correlation matrix similar to that shown in Table 6.13.

The techniques of elementary linkage analysis previously described were followed in isolating clusters of similar elections in terms of the Democratic percentage of the two party vote at the county level. Three such clusters were extracted from the matrix, the first of which consists of the elections of 1908, 1912, 1916, 1920, and 1924. Except for the 1920 election, the state supported the Democratic candidate in these elections. The prototype election for this cluster which represents the central tendency is 1916 in which Woodrow Wilson carried the state

TABLE 6.13 Matrix of Correlation Coefficient for Oklahoma Presidential Elections, 1908–1960

	1908	1912	1916	1920	1924	1928	1932	1936	1940	1944	1948	1952	1956	1960
1908		829	836	753	763	489	751	763	784	735	694	567	679	541
1912	829[a]		930	914	934	709	828	847	873	889	869	829	858	795
1916	836	930		895	907	589	859	884	863	867	855	736	818	680
1920	753	914	895		903	652	762	801	854	873	834	800	808	768
1924	763	934	907	903		773	869	850	893	904	907	855	890	803
1928	489	709	589	652	773		718	628	714	755	790	846	825	813
1932	751	828	859	762	869	718		888	845	847	885	730	868	710
1936	763	847	884	801	850	628	888		926	907	875	710	815	703
1940	784	873	863	854	893	714	845	926		978	937	850	899	822
1944	735	889	867	873	904	755	847	907	978		959	908	931	868
1948	694	869	855	834	907	790	885	875	937	959		909	956	875
1952	567	829	736	800	855	846	730	710	850	908	909		957	940
1956	679	858	818	808	890	825	868	815	899	931	956	957		922
1960	541	795	680	768	803	813	710	703	822	868	875	940	922	

[a] Decimal points have been eliminated.

by a very slim margin of 50.6 percent. Cluster two comprises the elections of 1932, 1936, 1940, 1944, and 1948. In each election a majority of Oklahoma voters supported the Democratic candidate. The prototype election for this cluster can be either 1940 or 1944. President Roosevelt received 57.4 percent and 55.5 percent respectively in the elections. The elections of 1948, 1952, 1956, and 1960 make up cluster three. These are four of the five presidential elections which the Republican party had won in Oklahoma. The prototype election for this cluster is 1952 in which the Democratic candidate, Adlai Stevenson, received only 45.4 percent of the vote.

These three clusters suggest that the Oklahoma electorate has experienced two periods of realignment since statehood in 1907. However, examination of the matrix reveals no negative correlations and only one less than .500. The uniformly high coefficients in each of the three clusters, except for 1908 and 1928, indicate that no dramatic and sudden realignment along the dimensions Key suggested has occurred. Certainly, 1928 was not a critical election. This is not unexpected since Oklahoma in 1928 did not have a high concentration of urban voters like that of Massachusetts. Actually, the 1928 election appears to be a deviant election. The trend of voting the 1928 election suggests did not emerge until 1952 and persisted through the 1960 election.

Elementary linkage analysis indicates that beginning in 1952 a significant number of Oklahoma counties began to show a gradually increasing Republican strength so that in the 1960s Oklahoma became a two party state. This immediately raises several questions about the demographic characteristics of these counties which may help to account for this change. Space does not permit elaboration of the research into recent Oklahoma social and political history that elementary linkage analysis generated. The point to be emphasized, however, is that elementary linkage analysis is not an end in itself. Rather, like the other quantitative techniques discussed in this book it can lead into more fertile areas of research.

Information Retrieval

The last area of computer applications requiring our attention is information retrieval. Of course retrieval is part of the data processing cycle in any analysis of computer legible data. However, computerized information retrieval is generally understood as procedures for storing, indexing, and retrieving textual material. Our discussion of computerized information retrieval in historical research will be suggestive rather than substantive, since historians are doing very little work in this rapidly expanding area.[60]

Automatic information retrieval of course utilizes information stored in machine readable form. This has given rise to a new kind of archive such as the Inter-University Consortium for Political Research which serves many scholars with varying research interests. The Consortium's holdings presently consist of complete county election returns since 1824 for the offices of President, Senator, Congressman, and governor, partial files of roll call votes for both houses of Congress since 1789, and complete published census data since 1790. Another kind of computerized data archive focuses on biographical data. The University of Pittsburgh Archive on Political Elites in Eastern Europe has biographical data on approximately 1200 persons who since 1945 have been members of the Central Committees, Secretaries, Councils of Minister, or Politburos of East Germany, Bulgaria, Czechoslovakia, Hungary, Poland, and Rumania. Lee Benson of the University of Pennsylvania and Theodore Rabb of Princeton University are currently working on a project which is both enormous and fascinating—computerization of biographical data of all Americans of any significant consequence. While the collection of such data for several hundred thousand Americans who have lived since 1700 is an enormous undertaking, it will be of incalculable value to scholars in general and historians in particular.

The present holdings of these data archives may be generally characterized as quantitative since most of the data files employ numerical codes. Computer technology has progressed to the point where natural language material may be handled as easily as numerical data. A basic approach followed in working with natural language material is keyword indexing. In this approach to computerized information retrieval a set of programs instructs a computer to scan a data file of natural language material for previously defined keywords, to index them in alphabetical order, and to print out these keywords along with the immediate context in which each occurs. There are two versions of keyword indexing information retrieval which are now operational. One is called KWIC for KeyWord-In-Context. The other is known as KWOC for KeyWord-Out-of-Context.

KWIC searches natural language material for keywords which are identified by reference either to a list of previously defined keywords or to a list of words not to be considered keywords.[61] The context of each keyword which may be retrieved is limited to 78 characters and spaces. After subtracting the number of characters in the keyword, the remaining space is divided and used to print out the text to the left and to the right of the keyword. The balance of space to the right is reserved for identifying the original source and exact location of the sentence or statement in which the keyword occurs. A KWIC index is permuted since each sentence or statement will appear as many times as the number of keywords it contains.

The usefulness of KWIC retrieval can be shown with several examples. The index to this book is a reproduction of the computer output generated by a version of the KWIC program.[62] Most historians would find a KWIC bibliographical search very helpful in locating published periodical or monograph information on a particular subject. Suppose that a KWIC index to the titles of all the articles published in the *American Historical Review,* the *Journal of American History,* and the *Journal of Southern History* were available. A researcher interested in slavery could locate all the articles in these journals on this subject by examining the keywords related to slavery.[63] The same researcher could use KWIC to search the MARC magnetic tapes for keywords relating to slavery and identify all the book titles dealing with slavery published since 1966.[64] A partial listing of a KWIC index to the *American Political Science Review* in Table 6.14 shows the format of output from KWIC retrieval.

TABLE 6.14. KWIK Listing in American Political Science Cumulative Index

TITLE	AUTHOR	I.D.#
KEYWORD	YEAR	
REPUBLICAN AND DEMOCRATIC NATIONAL COMMITTEES (PARTIES).=	PERSONNEL OF	SAYRE	WS32	1117
ADVISORY COMMITTEES IN BRITISH ADMINISTRATION.=		FAIRLI	JA26	826
ARLIAMENTARY ROLE OF JOINT STANDING COMMITTEES IN SWEDEN.=		ELDER	NC51	2069
CONFERENCE COMMITTEES IN THE NEBRASKA LEGISLATURE.=	THE P	BURDET	FL36	1365
SUB - COMMITTEES OF CONGRESS.=		FRENCH	BL15	315
MINISTRATION }.= PERMANENT ADVISORY COMMITTEES TO THE BRITISH GOVERNMENT DEPAR		PERKIN	JA40	1514
LEGISLATIVE INVESTIGATING COMMITTEES.=		MCCART	L 22	648
LEGISLATIVE INVESTIGATING COMMITTEES.=		BLAIR	JH24	718
N INTRODUCTION TO THE SENATE POLICY COMMITTEES.=	A	BONE	HA56	2299
ESTION TIME IN THE BRITISH HOUSE OF COMMONS (PARLIAMENT).=	QU	MCCULL	RW33	1202
A TEST FOR CABINET AUTOCRACY (OVER COMMONS IN BRITAIN).= WOMAN SUFFRAGE IN P		CLARK	E 17	423
).= AMENDMENTS IN HOUSE OF COMMONS PROCEDURE SINCE 1881 (PARLIAMENTS		PORRIT	E 08	033
Y OF THE PARLIAMENTS OF THE BRITISH COMMONWEALTH .=	THE COMMUNIT	HALL	HD42	1654
BRITISH EMPIRE (DEVELOPMENT OF THE COMMONWEALTH).= THE TREND WITHIN THE		BOGGS	TH15	357
BRITISH DOMINIONS AND NEUTRALITY (COMMONWEALTH).=	THE	CLOKIE	HM40	1540
THE BRITISH IMPERIAL CONFERENCE (COMMONWEALTH).=		SMELLI	KB27	857
FOREIGN POLICY AND THE DOMINIONS (COMMONWEALTH).=	BRITISH	DENNIS	AL22	643
L STATUS OF THE BRITISH DOMINIONS (COMMONWEALTH).=	INTERNATIONA	ALLIN	CD23	684
EREIGNTY OF THE BRITISH DOMINIONS (COMMONWEALTH).=	THE SOV	ELLIOT	WY30	1030
THE BRITISH COMMONWEALTH OF NATIONS.=		HALL	HD53	2190
IS THE BRITISH COMMONWEALTH WITHERING AWAY.=		WHEARE	KC50	2027
THE NATURE AND STRUCTURE OF THE COMMONWEALTH.=		WHEARE	KC53	2191
NALISM AND DEMOCRACY IN THE BRITISH COMMONWEALTH-- SOME GENERAL TRENDS.= NATIO		BRADY	A 53	2192
SEARCH).= INTER-AMERICAN SCHOLARLY COMMUNICATION IN THE HUMANITIES AND SOCIAL		BURKHA	F 60	2491
OCTRINAL CONFLICT (CATHOLICISM AND COMMUNISM).= DEVIATION CONTROL-- A STUDY		BRZEZI	Z 62	2547
SLAVIA').= HOW DIFFERENT IS TITO'S COMMUNISM (COMMENT ON 'THE COMMUNIST PART		DRAGNI	AN57	2335
REVOLUTIONARY COMMUNISM IN THE UNITED STATES.=		WATKIN	GS20	542
THE NEW STRATEGY OF INTERNATIONAL COMMUNISM.=		KAUTSK	JH55	2265
THE FRENCH PEASANT AND COMMUNISM.=		EHRMAN	HW52	2101
WESTERN EUROPEAN COMMUNISM-- A PROFILE.=		EINAUD	M 51	2061
STATE LEGISLATURES AND COMMUNISM-- THE CURRENT SCENE.=		PRENDE	WB50	2028
SOVIET POLICY TOWARD NATIONAL COMMUNISM-- THE LIMITS OF DIVERSITY.=		MORRIS	B 59	2435
ITY OF NATIONS AFTER WORLD WAR II.= COMMUNISM, NATIONALISM, AND THE GROWTH OF		SHOUP	P 62	2584
CONSTITUTIONALISM IN COMMUNIST CHINA.=		STEINE	HA55	2245
CURRENT 'MASS LINE' TACTICS IN COMMUNIST CHINA.=		STEINE	HA51	2066
NATIONALISM, AND THE GROWTH OF THE COMMUNIST COMMUNITY OF NATIONS AFTER WORLD		SHOUP	P 62	2584
WAR.= THE COMMUNIST DOCTRINE OF THE INEVITABILITY OF		BURIN	FS63	2604
E AS A FACTOR IN THE RECRUITMENT OF COMMUNIST LEADERSHIP.=	AG	HOLT	RT54	2220
D MODERATION.= THE TWENTIETH CPSU (COMMUNIST PARTY) CONGRESS-- A STUDY IN CA		KENNEY	CD56	2316
S ORGANIZATIONS IN MAINLAND CHINA (COMMUNIST PARTY).=	MAS	KUOCHU	C 54	2228
YOUTH UNDER DICTATORSHIP (RUSSIAN COMMUNIST PARTY).= THE KOMSOMOLS-- A STUD		FAINSO	M 51	2053
RECENT LITERATURE ON CHINESE COMMUNIST PARTY HISTORY.=		STEINE	HA52	2126
PPARATUS' OF THE CENTRAL COMMITTEE, COMMUNIST PARTY OF THE SOVIET UNION.= THE		NEMZER	L 50	2006
S OF A SOCIAL MOVEMENT.= THE COMMUNIST PARTY OF THE U.S.A.-- AN ANALYSI		MOORE	B 45	1757
THE COMMUNIST PARTY OF YUGOSLAVIA.=		NEAL	FW57	2334

TABLE 6.15.   Reproduction of a Partial Retrieval of 263 Roll Call
Votes in the Kansas State Senate 1957–1965.

```
MONEYS
    ASSESSMENT-AND-TAXATION COMMITTEE
    SENATE BILL 120-  AN ACT RELATING TO PUBLIC MONEYS-
    PROVIDING FOR THE PAYMENT OF INTEREST ON DEPOSITS OF FUNDS
    FROM THE STATE INACTIVE ACCOUNTS, AMENDING EXISTING K.S.A.
    75-2417 AND REPEALING SAID EXISTING SECTION.  THIS ACT WOULD
    CARRY OUT ANOTHER OF THE GOVERNOR'S REVENUE RECOMMENDATIONS
    BY INCREASING FROM ONE TO ONE AND ONE-HALF PERCENT THE
    INTEREST ON 80 PERCENT OF THE INACTIVE STATE FUNDS
    DEPOSITED IN KANSAS BANKS.  ESTIMATED REVENUE INCREASE IS
    $425,000.  REFERRED TO AND RECOMMENDED  FOR PASSAGE BY THE
    ASSESSMENT-AND-TAXATION COMMITTEE.  PASSED-- 31 YEAS,
    6 NAYS, 3 ABSENT OR NOT VOTING.
    JOURNAL OF THE KANSAS SENATE, 1965, PAGE 252.     COL. 26
    LEGISLATIVE SERVICE BULLETIN NO. 5, KANSAS STATE CHAMBER OF
    COMMERCE, PAGE 2.
MORTGAGE
    ADAMS CB            DAVIS JP            MIKESIC JM
    ROBERTS WL          VAN CLEAVE T
    HOUSE BILL 717-  AN ACT RELATED TO URBAN RENEWAL POWERS-
    PROVIDING FOR THE ISSUANCE OF BONDS TO FINANCE URBAN RENEWAL
    PROJECTS AND A MORTGAGE OF ANY SUCH URBAN RENEWAL PROJECT-
    AUTHORIZING A PLEDGE OF ALL EXCESS TAXES, AS A RESULT OF ANY
    INCREASE IN THE EXISTING VALUATION BECAUSE OF THE
    REDEVELOPMENT BY URBAN RENEWAL ACTIVITIES, TO THE PAYMENT
    OF PRINCIPAL AND INTEREST ON ANY BONDS ISSUED IN FINANCING
    SUCH PROJECTS AND A PLEDGE OF THE CREDIT OF THE MUNICIPALITY
    OR FEDERAL GOVERNMENT- PROVIDING FOR INVESTMENTS IN SUCH
    BONDS AND THE CONDITIONS THEREOF, AMENDING EXISTING K.S.A.
    17-4751 AND 17-4752 AND REPEALING SAID EXISTING SECTIONS.
    REFERRED TO AND RECOMMENDED FOR PASSAGE BY THE
    MUNICIPALITIES COMMITTEE.  ADVANCED TO THIRD READING BY
    EMERGENCY MOTION.  PASSED-- 25 YEAS, 7 NAYS, 8 ABSENT OR
    NOT VOTING.
    JOURNAL OF THE KANSAS SENATE, 1965, PAGE 386.     COL. 61
MOTELS
    HULTS DS
    SENATE BILL 68-  AN ACT RELATING TO HOTELS AND MOTELS AND TO
    HOTEL AND MOTEL KEEPERS, AND DEFINING THEIR LIABILITY FOR
    PROPERTY OF GUESTS.  REFERRED TO, AMENDED, RECOMMENDED FOR
    PASSAGE AS AMENDED BY JUDICIARY COMMITTEE.  PASSED--
    33 YEAS, 6 NAYS, 1 ABSENT OR NOT VOTING.
    JOURNAL OF THE KANSAS SENATE, 1959, PAGE 101.     COL. 22.
MOTOR
    IMMEL HM            JOSEPH CB           SMITH GS
    SENATE BILL 146-  AN ACT AUTHORIZING THE STATE DEPARTMENT OF
    ADMINISTRATION TO ESTABLISH AND MAINTAIN A MOTOR  POOL, AND
    PRESCRIBING THE METHOD OF OPERATION THEREOF.  REFERRED TO
    AND RECOMMENDED FOR PASSAGE BY  FEDERAL-AND-STATE-AFFAIRS
    COMMITTEE.  PASSED-- 23 YEAS, 6 NAYS, 11 ABSENT OR NOT
    VOTING.
    JOURNAL OF THE KANSAS SENATE, 1959, PAGE 122.     COL. 27.
```

From *Information Retrieval: Applications to Political Science*, by Kenneth Janda, copyright © 1968
by Bobbs-Merrill Company, Inc., reprinted by permission of the publishers.

Another version of keyword indexing is KWOC.[65] It differs from
KWIC in that the context is not restricted to 78 characters and the
keyword is set off to the left or "out-of-context" for easy scanning.
Since KWOC gives a full listing of entry information, it would be particularly useful in retrieving descriptive information about selected cate-

gories of roll call votes. Suppose a researcher wants to analyze the voting of U.S. Senators on roll call votes dealing with agriculture in the period between 1921 and 1933. During this period about 2000 roll call votes were recorded in the *Congressional Record* and the *Senate Journal.* There is no index to the *Journal* while the sheer bulk of the *Congressional Record* makes using its index difficult. The ICPR's holdings of Senate roll call votes for this period includes each Senator's vote along with a description of the vote taken directly from either the *Congressional Record* or the *Senate Journal.* Included in each description is the following: (1) author of the amendment or act; (2) date; (3) session of Congress; (4) purpose of the amendment, usually consisting of the exact wording of the amendment; (5) total yeas and nays along with a breakdown of the vote by party; (6) storage location of the votes of individual Senators; (7) source reference in either the *Journal* or the *Record.* With the KWOC system such keywords as agricultural, agriculture, cotton, crop, crops, farm, farmer, farmers, farms, and the like would retrieve for inspection all the descriptions of roll call votes pertaining to agriculture. On the basis of these descriptions roll calls could be selected for statistical analysis. A KWOC retrieval of the descriptions would resemble the one listed in Table 6.15.

The kinds of data searches historians can undertake with either KWIC or KWOC are determined by the questions being asked, the nature and scope of the data base, and the creative imagination of the user. An information retrieval system like KWIC or KWOC can not compensate for an inadequately formulated research design or fuzzy thinking. Information retrieval, like the other computer techniques examined in this chapter, is not an end in itself.

FOOTNOTES

1. *An Introduction to the Classification of Animals* (1869), quoted in Peter W. Hemingway "Multiple Agreement Analyis" (unpublished Ph.D. dissertation, Michigan State University, 1961), 1.

2. Roy E. SCHREIBER, "Studies in the Court of Wards, 1624–1635—The Computer as an Aid," *Computer Studies in the Humanities and Verbal Behavior,* I (January 1968), pp. 36–42.

3. SCHREIBER, "Studies in the Court of Wards," p. 37.

4. SCHREIBER, "Studies in the Court of Wards," p. 40.

5. SCHREIBER, "Studies in the Court of Wards," p. 41.

6. STEPHEN THERNSTROM, "The Historian and the Computer," in *Computers in Humanistic Research,* Edmund A. Bowles, ed., (Englewood Cliffs, N.J.: 1968), pp. 73–80.

7. THERNSTROM, "The Historian and the Computer," p. 79.

8. *The Alabama Review,* XIX (1966), pp. 243–76.

9. ALEXANDER, "The Basis of Alabama's Antebellum Two-Party System," p. 259.

10. These percentages are taken from W. Dean Burnham, *Presidential Ballots, 1836–1892* (Baltimore: 1955).

11. Program BMDO3R in W. J. Dixon, ed., *Biomedical Computer Programs* (Berkeley: 1965), pp. 258–75. This program was revised for use at the University of Alabama.

12. It should be noted that Professor Alexander in a paper on Correlation of Election Returns with Socio-Economic Information presented at the 1968 annual meeting of the American Association for State and Local History discussed some of the difficulties encountered in obtaining meaningful coefficients of partial correlation. He described the safeguards being used in his work with all of the counties of the United States for the 1840–1880 period to prevent use in multiple regression analysis of variables highly correlated with each other. The paper maintained that independent variables should not be used if bivariate correlations between the two is as high as .50, and that substantially higher correlation between any pair of independent variables will fatally distort the coefficient of partial correlation for each of them.

13. ALEXANDER, "The Basis of Alabama's Antebellum Two-Party System," p. 267.

14. ALEXANDER, p. 267.

15. *Comparative Studies in Society and History,* 5 (1963), pp. 134–63.

16. University of Pittsburgh Press, 1967.

17. Vanderbilt University Press, 1967.

18. The authors appreciate Professor Bogue's assistance in making available descriptions of the computer programs utilized in his study. The example used here is drawn from a paper Professor Bogue read at the Annual Meeting of the Organization of American Historians, Chicago, Illinois, 1967. The paper with minor revisions was subsequently published under the title of "Bloc and Party in the United States Senate: 1861–1863," *Civil War History,* 13 (1967), pp. 221–41.

19. BOGUE, "Bloc and Party," p. 222.

20. This program was written by a member of the staff of the University of Wisconsin Computer Center. A copy of the program, which is called BogueTab, may be obtained from Professor Bogue, Department of History, University of Wisconsin.

21. The high negative Q coefficient in this instance is treated as a positive coefficient.

22. For a discussion of cluster analysis see that section in this chapter.

23. While in Bogue's study the number of Senators and roll calls was not too large, manual posting is not recommended. Computer pro-

grams are available to do this. One such program will be described later in this section.

24. Bogue, "Bloc and Party," p. 241.

25. Bogue, "Bloc and Party," p. 224.

26. Alexander, *Sectional Stress and Party Strength,"* (Nashville: 1968), pp. 265–78.

27. Alexander, *Sectional Stress and Party Strength,* p. 10.

28. Alexander, *Sectional Stress and Party Strength,* p. 267.

29. Alexander offers a very useful guide for determining which cell is the "error" cell in the four-fold tables. See page 268.

30. Alexander, *Sectional Stress and Party Strength,* p. 11.

31. Alexander, *Sectional Stress and Party Strength,* p. 13.

32. For other examples consult the appropriate section in Chapter 8. A very useful description of Guttman scale analysis can be found in Lee Anderson, et al. *Legislative Roll Call Analysis* (Evanston: 1965).

33. S. Sidney Ulmer, "The Analysis of Behavior Patterns on the United States Supreme Court," *The Journal of Politics,* 22 (1960), pp. 629–53.

34. Ward H. Goodenough, "Some Applications of Guttman Scale Analysis to Ethnology and Culture Theory," *Southwestern Journal of Anthropology,* 19 (1963), pp. 235–50.

35. "A Scalogram Analysis of Political Development," *American Behavioral Scientist,* 9 (1966), pp. 33–36.

36. For a full description of the system along with applications see Philip J. Stone, Dexter C. Dunphy, Marshall S. Smith, and Daniel M. Ogilvie, *The General Inquirer* (Cambridge: 1966). The computer programs in this system are written for an IBM 7090/94 and an IBM 1401.

37. There are other dictionaries but they are not presently available for general use. The Stanford Political Dictionary is being rewritten in PL/I for use on the IBM 360 computers. The McPherson Lobbying Dictionary, which is described in *The General Inquirer,* is embedded in COMMIT, an obsolete programming language.

38. Stone, *The General Inquirer,* pp. 76–85.

39. A user's manual which fully describes how to set up each phase of the system is available from M.I.T. Press.

40. This conference was held at the University of Wisconsin, May 16–17, 1968.

41. "Rhetoric and Reality in the American Revolution," *The William and Mary Quarterly,* 3rd Ser., 22 (1966), pp. 3–32.

42. A 15 percent sample of Dickinson's work and Paine's work was used rather than the complete text. The procedures followed consisted of dividing the complete document into one inch sections and numbering them sequentially. A table of random numbers was then used to select 15 percent of the total number of one inch sections. If a section began in the middle of a sentence, an adjustment was made by moving up or down one or two lines in order to obtain a complete sentence thought. Since the single word is the primary unit

of analysis in the General Inquirer System, this was done primarily for the convenience of the researcher so that he could determine the immediate context of a tag without referring back to the original document.

43. Wood, "Rhetoric and Reality," p. 25.

44. See the provocative review "Literary Analysis with the Aid of the Computer: Review Symposium," *Computers and the Humanities*, 2 (1968), pp. 117–202. For another assessment see George Psathas, "The General Inquirer: Useful or Not?" *Computers and the Humanities*, 3 (1969), pp. 163–74. For research in the problem of word context of "word sense" see Philip Stone, "Improved Quality of Content Analysis Categories: Computerized Disambiguation Rules for High Frequency English Words" (A paper read at a National Conference on Content Analysis, The Annenberg School of Communication, University of Pennsylvania, November, 1967).

45. Dennis J. Arp, J. Philip Miller, and George Psathas, "An Introduction to the Inquirer II System of Content Analysis" (mimeographed, Washington University Computer Facilities, 1969); J. Philip Miller, ed., *Inquirer II Programmer's Guide* (Washington University Computer Facilities, 1969).

46. These questions are derived from Raymond B. Cattell, Malcolm A. Coulter, and Bien Jsujoika, "The Taxonmetric Recognition of Types and Functional Emergents," in *Handbook of Multivariate Experimental Psychology*, Raymond B. Cattel, ed. (Chicago: 1966), p. 293.

47. Cattell, *Handbook of Multivariate Experimental Psychology*, pp. 194, 294.

48. Raymond B. Cattell, *Description and Measurement of Personality* (New York: 1946), pp. 76–90.

49. Max Weber, "Ideal Types and Theory Construction," in *Readings in the Philosophy of the Social Sciences*, May Brodbeck, ed. (New York: 1968), pp. 496–507.

50. Raymond B. Cattell, *Factor Analysis* (New York: 1952), p. 30.

51. Cattell, *Factor Analysis*, p. 362.

52. The literature on this subject is voluminous. A good starting point is the essays in Patrick Gardiner, ed., *Theories of History* (New York: 1959).

53. Chapter 36 in Fred N. Kerlinger, *Foundations of Behavioral Research* (New York: 1964) provides one of the best brief expositions of this stage of factor analysis the authors have seen in print.

54. The source of this illustration is Charles M. Dollar, "Southern Senators and the Senate Farm Bloc, 1921–1925" (unpublished M.A. thesis, University of Kentucky, 1963).

55. David B. Truman, *The Congressional Party* (New York: 1959), pp. 45–50.

56. "Partisan Politics in the Continental Congress, 1778–1779: A Quantitative Approach" (unpublished paper).

57. "Elementary Linkage Analysis for Isolating Orthogonal and Oblique Types and Typical Relevancies," *Educational and Psychological*

Measurement, 17 (1957), pp. 207–29. Also see McQuitty, "Elementary Factor Analysis," *Psychological Reports,* 9 (1961), pp. 71–78.

58. CHARLES M. DOLLAR, "Innovation in Historical Research: A Computer Approach," *Computers and the Humanities,* 3 (1969), pp. 144–49.

59. "A Theory of Critical Elections," *Journal of Politics,* 17 (1955), pp. 3–18.

60. Our treatment of information retrieval draws heavily from Kenneth Janda, *Information Retrieval: Applications to Political Science* (Indianapolis: 1968). General trends in information retrieval as well as what historians are doing in this area are discussed in Dagmar Perman, ed., *Bibliography and the Historian: The Conference at Belmont* (Santa Barbara: 1968).

61. See the IBM General Information Manual E20-8091, "Keyword-In-Context (KWIC) Indexing," for specific information.

62. General information about preparation of the index for this book is given in the index.

63. If the titles to all of the articles published in 100 journals which relate directly or indirectly to history were in computer legible form, the KWIC search would disclose virtually all of the scholarly articles dealing with slavery which have been published in the United States.

64. MARC stands for MAchine Readable Cataloguing. Since November 1966 the Library of Congress has distributed to participating university libraries magnetic tapes which contain bibliographical information on English monograph titles processed by the Library of Congress. See Paul R. Reimers, "Progress of Automation" in *Bibliography and the Historian* (Santa Barbara: 1968) pp. 75–81.

65. JANDA, *Information Retrieval,* pp. 7–10, 72–73.

GUIDE TO RESOURCES OF VALUE IN QUANTITATIVE HISTORICAL RESEARCH

Chapter Seven

PART ONE: Guide to Published Data

A. *GENERAL GUIDES AND INTERNATIONAL SOURCES*

ALLEN, ROY GEORGE DOUGLAS, and J. EDWARD ELY, *International Trade Statistics*. New York: 1953.

The Americana Annual: An Encyclopedia of Current Events. New York: 1923. A continuing series. Only first year listed for serials still published.

Annuaire Colonial, annuaire agricole, commercial, et industriel des colonies francaises. Paris: 1888. A continuing series.

Appleton's Annual Cyclopaedia and Register of Important Events, 42 vols. New York: 1862–1903. The most useful of the yearbooks; good articles on states, countries, colonies, and religious groups.

Armaments Year-book, 15 vols. Geneva: 1924–1940.

AUFRICHT, HANS, *Guide to League of Nations Publications: A Bibliographical Survey of the Work of The League, 1920–1947*. New York: 1951.

BALL, JOYCE, *Foreign Statistical Documents: A Bibliography of General, International Trade and Agricultural Statistics*. Stanford, Calif.: 1967. Very convenient guide for publications of smaller countries after World War II.

BEACH, HARLAND, and ST. JOHN BURTON, eds., *World Statistics of Christian Missions*. New York: 1916. A major source, largely unused.

BLISS, WILLIAM D. P., *Encyclopedia of Social Reform*. New York: 1897. Unexplored.

Britannica Book of the Year. Chicago and London: 1938. A continuing series.

BROWN, EVERETT S., *Manual of Government Publications: United States and Foreign*. New York: 1950.

CANTRIL, HEDLEY, ed., *Public Opinion, 1935–46*. Princeton, N.J.: 1951. Important compilation of poll results.

Catalogue of the [Royal] Statistical Society, 3 vols. London: 1884–1886. Valuable; fine index.

The Columbia Lippincott Gazetteer of the World. New York: 1952, 1962. Good information on all cities, provinces, and so on, of world; editions of 1850, 1880, and 1893 are very useful.

Digest of Colonial Statistics. London: 1952. A continuing series.

Facts on File: A Weekly World News Digest With Cumulative Index. New York: 1941. A continuing series.

GREGORY, WINIFRED, ed., *List of the Serial Publications of Foreign Governments, 1815–1931*. New York: 1932. Massive basic source for series of statistical yearbooks and reports, for every country. Especially useful for late nineteenth, early twentieth century; lists holdings of major U.S. libraries.

JARVIS, R., "Official Trade and Revenue Statistics," *Economic History Review*, 17, (1964), pp. 43–62.

KEYFITZ, NATHAN, and WILHELM FLIEGER, *World Population: An Analysis of Vital Data*. Chicago: 1968. Massive compilation based on 500 censuses going back to 1780.

KELLER, HELEN R., *The Dictionary of Dates*, 2 vols. New York: 1934. Very detailed chronologies for all countries.

KOREN, JOHN, *History of Statistics, their Development and Progress in Many Countries*. New York: 1918. Pedestrian (but useful) reviews of the census and other statistical work of all major countries. Invaluable for nineteenth century when used with *List of Serial Publications* (1931).

LANGER, WILLIAM L., ed., *An Encyclopaedia of World History, Ancient, Medieval and Modern, Chronologically Arranged.* Boston: 1968.

LEES-SMITH, HASTINGS BERTRAND, *Encyclopaedia of the labour movement*, 3 vols. London: 1928.

London Bibliography of the Social Sciences, 4 vols. London: 1931. (Plus supplements.) Excellent guide to statistical books and journal articles, especially for late nineteenth, early twentieth century.

MAYR, GEORGE VON, *Statistik und Gesellschaftslehre*, 3 vols. Tubingen: 1895–1917. Classic.

MOOR, CAROL C., and WALDO CHAMBERLAIN, *How To Use United Nations Documents.* New York: 1952.

MORGENSTERN, OSCAR, *On the Accuracy of Economic Observation*, rev. ed. Princeton, N.J.: 1963. Important review of nature and types of errors in official economic sources, especially those of the United States.

MULHALL, MICHAEL GEORGE, *Dictionary of Statistics*, 4th ed., rev. London: 1899. List of books of reference; index to parts 1–2. Does not give authorities for statistics included; but very valuable compendium.

New Catholic Encyclopedia, 15 vols. New York: 1967.

New International Year Book: A Compendium of The World's Progress. New York: 1907. A continuing series.

Political Handbook and Atlas of the World: Parliaments, Parties and Press. New York: 1927. A continuing series.

Population Research Center (University of Texas): *International Population Census Bibliography*, 6 vols. Austin, Tex.: 1965–1967. Title listings, not annotated: vol. 1: Latin America and the Caribbean, 1965; vol. 2: Africa, 1965; vol. 3: Oceana, 1966; vol. 4: North America, 1966; vol. 5: Asia 1966; vol. 6: Europe, 1967. Very long and valuable, but of course incomplete, especially for early nineteenth century. Does not include provincial or local censuses.

REQUA, E. G., and J. STATHAM, *The Developing Nations: A Guide to Information Sources.* Detroit: 1965. Annotated bibliography focusing on economic development.

RICHARDSON, LEWIS F., *Statistics of Deadly Quarrels.* Chicago: 1960. Pioneering and very influential; includes more data than Pitirim Sorokin *Social and Cultural Dynamics*, vol. 3, New York: 1937.

ROACH, JOHN, *A Bibliography of Modern History.* Cambridge, England: 1968. A major guide, but only occasional annotations.

ROKKAN, STEIN, and J. MEYRIAT, eds., *International Guide to Election Statistics.* Paris: 1970. Five volumes planned; very valuable for bibliography and some data.

RUSSETT, BRUCE, H. R. ALKER, JR., KARL W. DEUTSCH, and HAROLD D. LASSWELL, *World Handbook of Social and Political Indicators*. New Haven, Conn.: 1964. Very useful compilation of 75 data sets; extremely important methodologically.

SILLS, DAVID L., ed., *International Encyclopedia of the Social Sciences*. 17 vols. New York: 1968. Consult the essays on the following topics: Census; Government Statistics; National Income; National Wealth; Vital Statistics; and Economic Data.

SLOCUM, ROBERT B., *Biographical Dictionaries and Related Works*. Detroit: 1967. 4829 biographical compilations cited and annotated for every country, and for occupational groups. Does not attempt full historical coverage, but very useful anyway.

SOROKIN, PITIRIM A., *Social and Cultural Dynamics*, 4 vols. New York: 1937–41. Major statistical study of wars and civil wars (v. 3); fascinating long-term collective biography in quantitative mode.

South America, *South American Handbook*. New York: 1921. A continuing series. Bureau of the American Republics, *Monthly Bulletin* (Washington: 1893–1910), good for statistics; its successor, the *Pan American Union Bulletin* (1910; continuing series) is less useful.

The Statesman's Year-Book: Statistical and Historical Annual of the States of the World. London and New York: 1864. A continuing series. The best and most convenient sources for national data; political, economic, military, education, government finances, etcetera, excellent bibliographic guides to primary sources.

STUDENSKI, PAUL, *A Study of the Income of Nations: Theory Measurement and Analysis, Past and Present*. New York: 1958. Summary of data for many countries historically, plus valuable bibliographic guide.

THOMAS, DANIEL, and LYNN CASE, eds., *Guide to the Diplomatic Archives of Western Europe*. Philadelphia: 1959. Comprehensive.

THORP, WILLARD, *Business Annals*. New York: 1926. Brief comments on the annual or quarterly state of commerce for major countries.

United Nations, *Demographic Yearbook. Annuaire demographique* New York: 1948. A continuing series of great value.

United Nations, *International Yearbook of Education*. 1939. A continuing series.

United Nations, *Statistical Yearbook*. New York: 1948. A continuing series.

U.S. Library of Congress, *National Censuses and Vital Statistics in Europe, 1918–39: An Annotated Bibliography*. Washington: 1948. *Supplement, 1940–48*. Washington: 1948.

U.S. Library of Congress, *Statistical Bulletins; An Annotated Bibliography of the General Statistical Bulletins of Major Political Subdivisions of the World,* prepared by Phyllis G. Carter. Washington: 1954. Good descriptions of contents of 1950 volumes. Major supplement to *List of Serial Publications.*

U.S. Library of Congress, *Statistical Yearbooks; An Annotated Bibliography of the General Statistical Yearbooks of Major Political Subdivisions of the World,* prepared by Phyllis G. Carter. Washington: 1953.

WEBB, AUGUSTUS DUNCAN, *New Dictionary of Statistics.* London, New York: 1911. A supplement for 1899–1909 to Mulhall's *Dictionary of Statistics.*

WILLCOX, WALTER F., ed., *International Migrations,* 2 vols. New York: 1929–1931. Basic compilation of data; good interpretations.

WINCHELL, CONSTANCE, M., *Guide to Reference Books,* 9th ed. Chicago: 1967. Basic bibliographical guide; periodic supplements.

Worldmark Encyclopedia of the Nations: A Practical Guide to the Geographic, Historical, Political, Social, and Economic Status of all Nations, Their International Relationships, and the United Nations System. New York: 1968.

WOYTINSKY, WLADIMIR S., *Die Welt in Zahlen,* 7 vols. Berlin: 1925–28. Very useful compilation of all sorts of economic, demographic, political, social, labor, military, financial data. Good for Europe, especially. Most of the time series begin during World War I.

WOYTINSKY, WLADIMIR S., and EMMA S. WOYTINSKY, *World Commerce and Governments: Trends and Outlook.* New York: 1955. Comprehensive.

WOYTINSKY, WLADIMIR S., and EMMA S. WOYTINSKY, *World Population and Production: Trends and Outlook.* New York: 1953. Comprehensive.

The Annual Register of World Events. London and New York: 1758. A continuing series. Good for British statistics and for record of events in leading nations.

League of Nations, *Statistical Yearbook of the League of Nations,* Geneva: 1931–1945.

B. *AMERICAN SOURCES*

1. General

American Baptist Convention Yearbook, 1950. Philadelphia: 1950. Annual. Continuation, under various titles, of a series that began with *Almanac and Baptists Register: 1841* (Philadelphia: 1840, annual). Being the most congregational body, the Baptists have the poorest statistics. For earlier sources see *Religion in Life* 25 (1956), p. 131.

American Jewish Yearbook. Philadelphia: 1899. A continuing series.

American Men of Science: A Biographical Directory, 11th ed., *The Social and Behavioral Sciences.* Tempe, Arizona: 1968. Cattell pioneered statistical collective biography of scientists; the latest edition is a good guide to colleagues; the companion *Dictionary of American Scholars,* (5th edition, 1969) covers the humanities and social sciences.

American Yearbook: A Record of Events and Progress. New York: 1910–1919, 1925–1950. Good on politics and scholarship.

AYER, N. W., *Ayer and Son's Directory of Newspapers and Periodicals.* Philadelphia, 1880, Annual. See also Rowells's *American Newspaper Directory,* 1869–1908. Basic source for circulation and party identification of newspapers; also useful for descriptive material on small towns and counties not in gazeteers.

BEATTIE, RONALD, "Sources of Statistics on Crime and Correction" *Journal of American Statistical Association,* 54 (1959), 582–592.

BODENSIECK, JULIUS, ed., *The Encyclopedia of the Lutheran Church,* 3 vols. Minneapolis: 1965.

BOECK, GEORGE A., "A Historical Note on the Uses of Census Returns" *Mid-America,* 44 (1962), 46–50.

Book of the States. Chicago: 1935. Biennial.

BOYD, ANNE MORRIS, *United States Government Publications,* 3d. ed., revised by Rae Elizabeth Rips. New York: 1949.

BURNHAM, W. DEAN, "Sources of Historical Election Data: A Preliminary Bibliography" (East Lansing, Mich.: 1967). Useful guide to almanacs, manuals.

BURR, NELSON ROLLIN, *A Critical Bibliography of Religion in America,* 2 vols. Princeton, N.J.: 1961. Very useful for locating statistical data; guides to collective biographies of ministers; missionary statistics located, etcetera.

CARROLL, HENRY K. *The Religious Forces of the United States.* New York: 1912. Analysis of census reports.

Collective biography: other sources
1. Local and county histories ("mugbooks") usually contain biographies of each subscriber. For a guide see Clarence Peterson, *Consolidated Bibliography of County Histories.* Baltimore: 1961.
2. State manuals and blue books give sketches of officeholders. Charles Press and Oliver Williams, *State Manuals, Blue Books* . . . Berkeley, Calif.: 1962.
3. Colleges collect, and most publish alumni directories, ranging from Sibley's *Harvard Graduates* (a major work of scholarship) to simple address lists. No convenient guide now exists, but see the shelf lists of major libraries under college heads.

4. Geneological societies track down all references to old families. See Constance Winchell's *Guide to Reference Books*, Chicago: 1967. Pp. 190–194.

CONGRESSIONAL QUARTERLY, *Congress and the Nation*, 2 vols. Washington: 1965–1969. Thoroughly reviews legislation 1945–1968, with key roll calls. For complete roll calls see *CQ Weekly Report*. Before 1945 try *Editorial Research Reports*, a continuing series from 1926.

Dictionary of American Biography, Published under the auspices of the American Council of Learned Societies. New York: 1928–1937. 14,000 biographies, many long, all by scholars; guide to obituaries and other sources.

EASTERLIN, R. A., *Interregional Differences in Per Capita Income, Population, and Total Income, 1840–1950*. In National Bureau of Economic Research, Conference on Research in Income and Wealth, *Trends in the American Economy in the Nineteenth Century*. Princeton, N.J.: 1960. Pp. 73–140. (Rest of book also very important.)

Economic Almanac. New York: 1940. A handbook of useful facts about business, labor and government in the United States and other areas. A continuing series.

Encyclopedia of Southern Baptists, 2 vols. Nashville: 1958.

The Episcopal Church Annual. New York: 1830. Annual.

FOLGER, JOHN K. and CHARLES NAM, *Education of the American Population*. Washington: 1967. Important compilation of statistics.

GAUSTAD, EDWIN SCOTT, *Historical Atlas of Religion in America*. New York: 1962. Graphic-statistical presentation of membership growth and church location. Good bibliographic suggestions. Cartography inferior to magnificent maps in 1890. U.S. Census, volume on churches. See also 1906, 1916, 1926, and 1936 U.S. Census of Churches.

HASSE, ADELAIDE ROSALIE, *Index of economic material in documents of the states of the U.S.* Washington: 1907–1922, 13 vols. in 16. Volumes issued are:
California, 1849–1904. 1908.
Delaware, 1789–1904. 1910.
Illinois, 1809–1904. 1909.
Kentucky, 1792–1904. 1910.
Maine, 1820–1904. 1907.
Massachusetts, 1789–1904. 1908.
New Hampshire, 1789–1904. 1907.
New Jersey, 1789–1904. 1914.
New York, 1789–1904. 1907.
Ohio, 1787–1904. 1912.
Pennsylvania, 1790–1904. 1919–22. 3 vols.
Rhode Island, 1789–1904. 1908.

Vermont, 1789–1904. 1907.
Omits reports of bureaus of labor before 1902, as these are covered
in the *Index to All Reports Issued by Bureaus of Labor Statistics . . . to March 1902,* issued by the U.S. Bureau of Labor.

HOLLEB, DORIS B. *Social and Economic Information for Urban Planning,* 2 vols. Chicago: 1969. Very thorough review of quality and
extent of contemporary data sources, with annotated bibliography
that is especially strong on little-known governmental and other
official reports.

IRELAND, NORMA O. *Local Indexes in American Libraries: A Union
List of Unpublished Indexes.* Boston: 1947. Comprehensive guide
to numerous card files maintained by public libraries. Important
for local history, indexes, censuses, lists of public officials, collective biography, indexes of newspapers. For example, the Connecticut State Library has 655,000 cards on state residents.

Lutheran World Almanac. New York: 1921. A continuing series.

MACHLUP, FRITZ, *The Production and Distribution of Knowledge
in the United States.* Princeton, N.J.: 1962. Basic.

Martindale-Hubble Law Directory. New York: 1968. Annual lists
and ratings of lawyers in U.S. and Canada.

MCPHERSON, EDWARD, ed., *A Hand-Book of Politics.* Washington:
1868–1894, biennial. Handy for roll calls.

Methodist Church, *General Minutes of the Annual Conferences.*
1773. A continuing series. Place varies. Magnificently detailed
statistics, by church, especially good for nineteenth century.

Monthly Labor Review. Washington, 1915. A continuing series.
Contains special reports in the field of labor. Statistics cover employment, labor turnover, earnings and hours, work stoppages,
prices, and cost of living, etcetera. Each issue also contains a
bibliography of recent labor literature.

MONROE, PAUL, *Cyclopedia of Education,* 5 vols. New York:
1911–1913. Old but still the most comprehensive work. Good statistics, for example, on distribution of college alumni by
occupation.

MORRIS, RICHARD B., ed., *Encyclopedia of American History.* New
York: 1970. Very helpful.

Municipal Yearbook. Chicago: 1934. Annual.

National Council of the Churches of Christ in the U.S.A., Bureau
of Research and Survey, *Churches and church Membership in
The U.S.; an Enumeration and Analysis by counties, states, and
regions.* New York: 1956–1958. Series of pamphlets giving *white*
church membership by denomination by county, based on ques-
tionnaires sent out in 1952. Not especially reliable, but only source
since Census was forced to drop religious census of churches in
1936.

Negro Yearbook, 1912–1952, 11 vols. Tuskegee, Ala.

NORTH, SIMON NEWTON DEXTER, *History and Present Condition of the Newspaper and Periodical press of the U.S., with a Catalogue of the Publications of the Census year.* Washington: 1884 (U.S. Census office, 10th Census, vol. 8). Maps and statistical tables.

PAULLIN, CHARLES O. and JOHN K. WRIGHT, eds., *Atlas of the Historical Geography of the United States.* Washington: 1932. Includes statistical maps.

PERLOFF, HARVEY, et al., *Regions, Resources and Economic Growth.* Baltimore: 1960.

Poor's Register of Corporations, Directors, and Executives, U.S. and Canada. New York: 1928. A continuing series.

POTTER, NEAL, and FRANCIS T. CHRISTY, JR., *Trends in Natural Resource Commodities: Statistics of Prices, Output, Consumption, Foreign Trade and Employment in the U.S., 1870–1957.* Baltimore: 1962.

PRESS, CHARLES and WILLIAMS, OLIVER, *State Manuals, Blue Books, and Election Results.* Berkeley: 1962.

PRESSLY, THOMAS J., and WILLIAM H. SCOFIELD, eds., *Real Estate Values in the United States.* Seattle: 1959.

REYNOLDS, LLOYD GEORGE and KILLINGSWORTH, CHARLES C., *Trade Union Publications: The Official Journals, Convention Proceedings, and Constitutions of International Unions and Federations, 1850–1941,* 3 vols. Baltimore: 1944–1945. Very useful guide to history of small unions.

Southern Baptist Convention, *Executive Commission Annual.* Nashville, Tenn.: 1847. Annual statistical compilations.

Statistics Sources, Paul Wasserman et al., eds. Detroit: 1962. Mostly for current series.

SUTHERLAND, STELLA, "Colonial Statistics," *Explorations in Entrepreneurial History,* 5 (1967), pp. 58–107. Thorough discussion of American data, based on *Historical Statistics of the United States* (1960 ed.)

United Church of Christ Yearbook. Philadelphia: 1962. Continues *Congregational Yearbook* (1853–1961).

United Presbyterian Church in the United States of America, General Assembly, *Minutes of the General Assembly* (Philadelphia). Title varies. Basic statistical source.

U.S. Board of Governors of the Federal Reserve System, *All-Bank Statistics, United States, 1896–1955.* Washington: 1959.

U.S. Bureau of the Budget. Office of Statistical Standards, *Statistical Services of the United States Government,* Revised ed. Washington: 1963.

U.S. Bureau of the Census, *Catalog of U.S. Census Publications, 1790–1945.* Washington: 1950. With annual supplements.

U.S. Bureau of the Census, *Compendium of the Census*. Washington: decennial, 1850–1920. Useful.

U.S. Bureau of the Census, *Congressional District Data Book*. Washington: 1961. Biennial.

U.S. Bureau of the Census, *County and City Data Book*. Washington: 1967 ed. Very valuable.

U.S. Bureau of the Census, *Negro Population, 1790–1915*. Washington: 1918.

U.S. Bureau of the Census, *Religious Bodies*. Washington: 1890, 1906, 1916, 1926, 1936.

U.S. Bureau of the Census, *Statistical Atlas of the United States*. Washington. For 1870, 1890, 1900, and 1910 censuses.

U.S. Bureau of the Census, *Statistical Abstract of the United States*. Washington: 1878. Basic. Better every year. Annual.

U.S. Bureau of the Census, *Historical Statistics of the United States, 1789–1945: A Supplement to the Statistical Abstract of the United States*. Washington: 1949. First edition of the basic sourcebook, *Historical Statistics of the United States from Colonial Times to 1957*. Washington: 1960. The 1960 edition should be on every historian's shelf. A 1965 *Supplement* continues the series to 1962.

U.S. Bureau of Labor Statistics, *Employment and Earning Statistics for the United States, 1907–66*. Washington, 1966.

U.S. Bureau of Labor Statistics, *Handbook of Labor Statistics*, Washington: 1927–1951.

U.S. Bureau of Labor Statistics, *History of Wages in the United States from Colonial Times to 1928*. Washington: 1934.

U.S. Congress, *Biographical Directory of the American Congress, 1774–1961*. Washington: 1961. Basic.

U.S. Department of Agriculture, *Yearbook of Agriculture*. Washington: 1894. Annual.

U.S. Library of Congress, Census Library Project, *Catalog of United States Census Publications, 1790–1945*. Edited by Henry J. Dubester, chief, Census Library Project. Washington: 1950. *The History and Growth of the United States Census*. Washington: 1900. U.S. Office of the Census. *A Century of Population Growth from the First Census . . . to the Twelfth, 1790–1900*. Washington: 1909, its *Circular of Information concerning census publications, 1790–1916*. January 1, 1917, its *Topical Index of Population Census Reports, 1900–1930*. 1934, its *Periodic and Special Reports on Population, 1930–1939*.

U.S. Library of Congress, Census Library Project, *State Censuses: An Annotated Bibliography of Censuses of Population Taken after the year 1790 by States and Territories of the United States*, edited by Henry J. Dubester. Washington: 1948. Very important guide to state censuses, usually taken midway between Federal censuses. New York (1845–1885), Massachusetts, Michigan, Iowa, Rhode Island, Wisconsin, are the most important. Contain minor civil division data not in U.S. Census, and asks different questions. Religious affiliation asked in Rhode Island, Iowa, South Dakota.

U.S. National Archives, *Federal Population Censuses, 1840–90; a Price List of Microfilm Copies of the Original Schedules.* Washington: 1963. (National Archives publ. no. 55–7.) Purchasing guide; all are available, by county at about $8 per reel.

U.S. Office of Education, *List of Publications, 1867–1910.* Washington: 1910. Reprinted, 1940. Educational statistics annually, plus good college data, community and state surveys, Negro education, and so forth. Before 1867, best source is *American Journal of Education* (edited by Henry Barnard).

U.S. Office of Education, *List of Publications, 1910–1936.* Washington. 1937. *1937–1959,* 1960.

U.S. Superintendent of Documents, *Checklist of U.S. Public Documents, 1789–1909,* 3rd ed. revised. Washington: 1911. The immense amount of statistical data collected by Congress is very difficult to locate. This guide serves as a first index to Congressional reports, as well as executive department reports. For later material, see the *Monthly Catalog* of U.S. Government Publications (1895).

Who Was Who in America: A Companion Biographical Reference Work to Who's Who in America. Chicago: 1942. Very useful.

Who Was Who in American Historical Volume, 1607–1896, a companion volume of *Who's Who in American History.* Chicago: 1963. Very short biographies; more names than *DAB.*

WILCOX, JEROME KEAR, *Manual on the Use of State Publications.* Chicago: 1940. Immense quantities of statistical data exist, but location is haphazard, and only the largest libraries have good collections; local and city documents are also useful for taxation, police, social, educational data, but no convenient guide exists.

WILSON, J. G. and JOHN FISKE, eds., *Appleton's Cyclopaedia of American Biography,* 7 vols. New York: 1887–1900. Short, uncritical biographies; useful mainly as supplement to *DAB.*

Yearbook of American Churches. New York: Published by the National Council of the Churches of Christ in the U.S.A., 1916. Publishes data of dubious reliability, with few state series. See Benson Y. Landis, in *Journal American Statistical Association* (1958).

2. Sources of American Election Statistics

    a. Machine readable: The Inter-University Consortium for Political Research (Ann Arbor, Mich.) has available on cards or computer tape all existing county returns for presidential, gubernatorial, and senatorial elections since 1824. A 10 volume printout of the ICPR election archives will be published by Wiley-Interscience in the early 1970s.

b. Compilations of Presidential Votes by County:

BURNHAM, WALTER DEAN, *Presidential Ballots, 1836–1892*. Baltimore: 1955. Percentages are not given.

PETERSEN, SVEND, *A Statistical History of the American Presidential Elections*. New York: 1963. Percentages for states.

ROBINSON, EDGAR E., *The Presidential Vote, 1896–1932*. Stanford, Calif.: 1934. Percentages are not given.

c. Other Major Compilations:

SCAMMON, RICHARD, *America Votes*. Washington: 1958. County and big city ward returns along with percentages for presidential, gubernatorial, and senatorial races. District returns for congressional races are also given. Biennial since 1958.

HEARD, ALEXANDER, and DONALD STRONG, *Southern Primaries and Elections, 1910–1949*. (University of Alabama: 1950.) Data for major primaries in 11 states. Some percentages are given.

Republican National Committee, *Summary Report*. Biennial since 1938.

ABC, CBS, and NBC television news divisions have extensive computer archives of representative precinct data for primary and general elections during the 1960s.

d. Almanacs: Useful for county and sometimes precinct returns.

New York *Tribune Almanac* (1838–1914).

Chicago *Daily News Almanac* (1885–1938).

New York *World Almanac* is unreliable and should not be used.

e. Compilations of Gubernatorial and/or Senatorial Returns by state and editor; with percentages.

Alabama. Lewy Dorman (1850–1860).

Arizona. Bruce Mason (1911–1960).

California. Eugene Lee (1928–1960).

Illinois. T. C. Pease (1818–1848); Samuel Gove (1900–1958).

Indiana. Dorothy Riker (1816–1851); Robert Pitchell (1852–1956).

Kansas. Clarence Hein (1859–1956).

Kentucky. Jasper Shannon (1824–1948); Malcolm Jewell (1920–1960).

Maryland. Evelyn Whitworth (1934–1958).

Michigan. John P. White (1929–1956).

Montana. Ellis Waldron (1864–1956).

Nebraska. *Nebraska Blue Book* (1918 ed., 1854–1916).

New York City. *Manual of the Corporation of the City of New York* (1871). Pp. 726–759 gives all city returns for mayor, president, governor, and secretary of state by ward with percentages for 1834–1869.

Oklahoma. Oliver Benson (1907–1964).
South Dakota. Alan Clem (1928–1960).
Texas. David Olson (1948–1963).
West Virginia. William Ross (1916–1952).
Wisconsin. James Donoghue (1848–1960).

C. OTHER COUNTRIES

Allgemeine deutsche Biographie; hrsg. durch die Historische Commission bei der K. Akademie der Wissenschaften, 56 vols. Leipzig: 1875–1912.

Argentina. *Anuario Estadistico.* (1892–1914, 1941, 1948). Annual.

Australia. Bureau of Census and Statistics, *Official Yearbook of the Commonwealth of Australia.* (1908). Annual.

Austria. *Österreichisches Jahrbuch.* Vienna: 1919. A continuing series.

Belgium. Institut National de Statistique. *Annuarie statistique de la Belgique.* (1870). Annual.

Brazil. *Anúario estatístico do Brasil.* (1908–1912, 1936). Annual.

BUTLER, DAVID, and JENNIE FREEMAN, *British Political Facts, 1900–1967.* New York: 1968.

Canada. *The Canada Year Book.* (1905). Continues Department of Agriculture statistical *Yearbook of Canada* (1886–1904). Annual.

CHEN, NAI-RUENN, *Chinese Economic Statistics.* Chicago: 1966. Extensive handbook of mainland data.

Chile. *Anuario estadistico* (1860–1888, 1909). Annual.

China Year Book. (1912–1939). Basic source; place varies.

CLARK, COLIN, *The Conditions of Economic Progress,* 3rd ed. London: 1957.

CLARK, G. N., *Guide to British Commercial Statistics, 1696–1782,* London: 1938. Standard.

California, University of at Los Angeles. Committee on Latin American Studies. *Statistical Abstract of Latin America.* Los Angeles: 1955–1956. Biennial, starting with 1964–1965; previously annual.

DAMPEIRRE, JACQUES DE, *Les publications officielles des pouvoirs publics: Ettude critique et administrative.* Paris: 1942. Detailed historical guide to French government publications.

DEWHURST, J. FREDERIC, et al., *Europe's Needs and Resources: Trends and Prospects in Eighteen Countries.* New York: 1961.

Dictionnaire de biographie francaise. Paris: 1933. In progress; A to D finished. Major French biographical dictionary.

Dictionnaire des parlementaires francais: Notices Biographiques sur les ministres, senateurs et députés francais de 1889 à 1940. Paris: 1960. In progress.

Dizionario biografico degli italiani. (Redazione, direttorei Alberto M. Ghisalberti) Rome: 1960. In progress.

DUBESTER, HENRY J., *National Censuses and Vital Statistics in Europe: 1918–1929.* Washington: 1948. Very detailed annotations. Indicates bilingual volumes 1940–1948.

DUBESTER, HENRY J., *Population Censuses and other Official Demographic Statistics of British Africa.* Washington: 1950. Nineteenth century material cited for Basutoland, Gambia, Gold Coast, Kenya, Mauritius, Lagos, St. Helena, Seychelles, Sierra Leone, Rhodesia, Cape of Good Hope, Natal, Transvaal, Orange Free State, and German and British Colonial Offices.

Egypt. *Annuaire Statistique.* Cairo: 1909. Annual.

FORD, PERCY, and GRACE FORD, eds., *Guide to Parliamentary Papers.* Oxford: 1956.

FORD, PERCY and GRACE FORD, *Select List of British Parliamentary Papers, 1833–1899.* Oxford: 1969.

FORD, PERCY and GRACE FORD, *A Breviate of Parliamentary Papers, 1900–1916; the Foundation of the Welfare State.* Oxford: 1957.

France. Institut National de la Statistique et des Etudes Economiques. *Annuaire statistique de la France (1877).* Annual.

Germany (Federal Republic): Statistisches Bundesamt, Wiesbadan. *Statistisches Jahrbuch für die Bundesrepublik Deutschland.* (1952). A continuing series. Continues the *Statistisches Jahrbuch für das Deutsche Reich,* 57 vols. (1880–1938). For the territory under the Federal Government of Germany. Has an added section giving international statistics.

GILLE, BERTRAND, *Les Sources statistiques de l'histoire de France; des Enquetes du XVIIe Siècle à 1870.* Paris: 1964.

Great Britain: Central Statistical Office. *Annual abstract of statistics.* (1854). Annual.

Great Britain. Colonial Office. *Colonial Reports.* London: 1891–1940.

Great Britain. Foreign Office, Handbooks prepared under the direction of the Historical Section of the Foreign Office, ed. by Sir G. W. Prothero. (162 parts in 25 vols.) London: 1920.

Great Britain. Interdepartmental Committee on Social and Economic Research. *Census Reports of Great Britain, 1801–1931.* London: 1951.

Great Britain. Interdepartmental Committee on Social and Economic Research. *Guides to Official Sources.* London: 1953.

Hungary; Központi Statisztikai Hivatal. *Statisztikai évkönyv. Statistical Yearbook, 1949–55.* Budapest: 1957. Published in Hungarian and English. Supersedes the Office's Magyar statisztiki évkönyv. *Statistisches Jahrbuch.* vol. 1–20, 1871–1890; *Annuaire statistique.* (Nouv. cours, 1893–1941.) Annual.

Institut International de Statistique. Office Permanent. *Statistique internationale des grandes villes,* 3 vols. Le Havre: 1927–1940.

Institute of Electoral Research, London. *Parliaments and Electoral Systems, A World Handbook.* London: 1962.

Instituto Centrale di Statistica *Compendio del Statistiche Elettorali Italiane dal 1848 al 1934,* vol. 1. Rome: 1964. Data on registration, eligible electorate and votes cast, and results, by provinces and large cities.

Instituto Centrale di Statistica, "Le Rilevazioni Statistiche in Italia dal 1861 al 1956," *Annali di Statistica* (Series 8, Vol. 5, 6, Rome). On organization and method of official statistics.

Instituto Geográfico y Estadístico, *Resensa Geográfica y Estadística de España,* 1888; 2nd ed. Rome: 1912. Useful statistical compendium for nineteenth century Spain, includes political, economic and social data, and territorial statistics. Maps and time series.

Instituto Nacional de Estadistica, *Principales Actividades de la Vida Española en la Primera Mitad del Siglo XX. Síntesis Estadística.* Statistical series, 1900–1950. Spain also published a statistical yearbook, 1860–1867, 1912–1935, 1943. Annual.

Inter-American Statistical Institute. *Bibliography of Selected Statistical Sources of the American Nations. Bibliografía de fuentes estadísticas escogidas de las naciones Americanas.* Washington: 1945.

International Labor Office, Geneva, *Yearbook of Labour Statistics.* Geneva: 1930. Annual.

Italy. Istituto Centrale di Statistica. *Annuario statistico italiano.* Rome: 1878. Annual.

Italy. Istituto Centrale di Statistica, *Sommario di statistiche storiche italiane, 1861–1955.* Rome: 1958.

Japan. Bureau of General Statistics. *Résumé statistique de l'Empire du Japon.* Tokyo: 1887–1940. Continued by *Japan Statistical Yearbook.* Tokyo: 1949. Annual.

Japan. Ministry of Finance, *Financial and Economic Annual of Japan.* Tokyo: 1901–1940.

KENDALL, MAURICE G. *The Sources and Nature of the Statistics of the United Kingdom,* ed. for the Council of the Royal Statistical Society. London: 1952–1957.

L'Année politique. 1st series, 1874–1905; 2nd series, 1944–1945. Paris: 1876–1906, 1945. Chronological review of major political, diplomatic, economic and social events in France and French overseas territories. An invaluable guide for French politics and foreign affairs. Abundant documentary material and French electoral statistics. Good index.

Library of Congress, *Biographical Sources for Foreign Countries.* Washington: 1944–1945, 3 parts. Part 1—general; very useful guide to biographical sources of men living in 1930s and earlier. Part 2—(1945) Germany and Austria—most convenient guide for collective biography sources. Part 3—(1945) the Philippines.

MADOZ, PASCUAL, *Diccionario Geografico-Estadistico-Historico de Espana y Sus Posesiones de Ultramar.* Monumental compendium. Madrid: 1845. See also S. Minano, *Diccionario Geografico-Estadistico de Espana y Portugal* (1826).

MAYR, GEORGE VON, ed., *Die Statistik in Deutschland nach ihrem heutigen stand,* 2 vols. Munich: 1911. Thorough review of sources and research in German tradition.

MELON, ARMADE, "Los Censos de Poblacion en Espana, 1857–1940," *Estudios Geograficos* 43: (1951), 303. Guide to contents.

Mexico. *Anuario Estadistico de los Estados Unidos Mexicanos* (1893–1907, 1930, 1938). Annual.

MITCHELL, BRIAN R., *Abstract of British Historical Statistics.* Cambridge: 1962. Basic.

MITCHELL, B. R. and BOEHM, K., *British Parliamentary Election Results, 1950–1964.* Cambridge: 1966.

MONTENEGRO, TULO H. "Bibliografia Anotada de las Principales Fuentes de Estadisticas Sobre America Latina," in Henry E. Adams (ed.) *Handbook of Latin American Studies: Number 29, Social Sciences.* Gainesville, Fla.: 1967, pp. 613–639. Annotated guide to recent serial publications and census reports.

NEALE, EDWARD P., *Guide to New Zealand Official Statistics,* 3rd ed. Christchurch: 1955.

Netherlands. Centraal Bureau voor de Statistiek, *Jaarcifers voor Nederland. Statistical yearbook of the Netherlands.* Gravenhage: 1884. Text in Dutch and French, 1884–1939; in Dutch and German, 1940–41/42; in Dutch and English, 1943/46. Annual.

New Zealand, Department of Statistics, *New Zealand Official Yearbook.* Wellington: 1892.

Norway. Statistiske Sentralbyrå, *Statistisk årbok for Norge,* (*1880*). A continuing series.

Parliamentary Handbook of the Commonwealth of Australia. Canberra: 1915. A continuing series.

Philippines (Republic). Bureau of the Census and Statistics, *Philippine Statistics, 1903–1959.* Manila: 1960.

POWICKE, FREDERICK M. and E. B. FRYDE, *Handbook of British Chronology,* 2nd ed. London: 1961.

REMOND., RENE et al., *Atlas Historique de la France Contemporaine: 1800–1965*. Paris: 1966. Important statistical maps.

ROBERT, ADOLPHE, EDGAR BOURLOTON and GASTON COUGNY, *Dictionnaire des parlementaires francais, comprenant tous les membres des assemblees francaises et tous les ministres francais depuis le 1ᵉʳ mai 1789 jusqu'au 1ᵉʳ mai 1889, avec leurs noms, etat civil, etats de services, actes politiques, votes parlementaires, etc.* Paris: 1891. 5 vols.

Rumania. Institutul Central de Statisticâ. *Anuarul statistic al României. Annuaire statistique de la Roumanie*, 1902–37/38. Text and tables in Rumanian and French.

SBROJAVACCA, L. "Della Finanze della Amministrazioni Locali in Alcuni Stati Europei," *Bulletin de l'Institute Internationale de Statistique*, I (part 2) (1887), pp. 199–299.

SCHANZER, CARLO, "Notizie sull' Ordinamento del Polere Legislativo: E sulle Elezione Politiche nei Principali Stati d'Europa," *Bulletin de l'Institute international de statistique*, II (part 2) (1887) 244–321. Data on turnout, registration, party division of elections of the 1870s and 1880s in 12 European countries.

South Africa. Office of Census and Statistics, *Official Yearbook of the Union and of Basutoland, Bechuanaland Protectorate and Swaziland*. Pretoria: 1917. Bienniel.

Spain. Instituto Nacional de Estadistica. *Anuario estadistico de Espana* (1912). Annual.

Sweden. Statistika Centralbyran. *Statistisk arbok for Sverige. Statistical yearbook of Sweden*. Stockholm: 1914. Annual.

———, *Historical Statistics of Sweden. I: Population, 1720–1950*. Stockholm: 1955.

Switzerland, Statistisches Amt. *Statistisches Jahrbuch der Schweiz. Annuaire statistique de la Suisse*. Bern: 1891. Annual.

TAEUBER, IRENE E., *General Censuses and Vital Statistics in the Americas*. Washington: 1943. Census project, similar to Dubester.

THOMPSON, DAVID M., "1851 Religious Census: Problems and Possibilities," *Victorian Studies* 11 (1967), pp. 87–97.

Times of India Directory and Yearbook Including Who's Who. Bombay, London: 1914. Annual.

USSR. Tsentral' noe Statisticheskoe Upravlenie. *Statistical Handbook of the USSR;* with introduction, additional tables and annotations by Harry Schwartz. Moscow: 1956; New York: 1957.

U.S. Bureau of the Census. Summaries of *Biostatistics: Maps and Charts, Population, Natality, and Mortality Statistics*. Prepared in cooperation with Office of the Coordinator of Inter-American Affairs, 17 vols. Washington: 1944–1945.

U.S. Bureau of the Census. *The Population and Manpower of China: An Annotated Bibliography,* by Foreign Manpower Research Office. Washington: 1958.

URQUHART, M. C., BUCKLEY, K. A., eds., *Historical Statistics of Canada.* New York: 1965. Basic.

WEBB, H., *Research in Japanese Sources: A Guide.* New York: 1965.

PART TWO: Social Science Machine Readable Data

Archives in the United States

Archive on Comparative Political Elites. University of Oregon, Eugene. Data on background and recruitment of political elites in the United States, Israel, and East and West Germany.

Archive on Political Elites in Eastern Europe. University of Pittsburgh, Pittsburgh. Biographical information on 1200 persons who, since 1945, have been members of the Central Committees, Secretaries, Councils of Minister, or Politburos of East Germany, Bulgaria, Czechoslovakia, Hungary, Poland, and Rumania.

Bureau of Labor Statistics. United States Department of Labor. The vast holdings of this archive include surveys of labor turnover in industry, employment payroll and hours, consumer expenditures, and estimates of labor force characteristics.

Bureau of Applied Social Research. Columbia University, New York. Sample surveys and panel studies which cover such subjects as health, welfare, occupations and professions, mass communications, politics, and education.

Council for Inter-Societal Studies. Northwestern University, Evanston, Ill. A variety of cross-cultural data, including some sample survey studies. In addition, state legislative data for Indiana from 1939–1961 are in the archive.

Data Archive, Graduate School of Industrial Administration. Carnegie-Mellon University, Pittsburgh. Ecological data, survey data, and election statistics from French cantons.

International Data Library and Reference Service. Survey Research Center, University of California, Berkeley. Sample surveys dealing primarily with politics and public opinion in Asia and Latin America.

International Development Data Bank. Michigan State University, East Lansing. Sample surveys and panel studies principally conducted in Latin America, Africa, and Asia.

Inter-University Consortium for Political Research. University of Michigan, Ann Arbor. Major archive in the United States for machine-readable data. Holdings center on the United States with an emphasis on survey data, and complete files of election statistics, Congressional roll calls, and census data. There are some cross-national survey studies and comparative international data. Now adding historical data from France.

Louis Harris Political Data Center. University of North Carolina, Chapel Hill. Public opinion surveys dealing with politics conducted by Louis Harris and Associates, Inc. in the United States since 1956.

M.I.T. Social Science Data Bank. Massachusetts Institute of Technology, Cambridge, Mass. Sample surveys conducted throughout the world which cover politics, social behavior, public opinion, and urban affairs.

National Opinion Research Center. University of Chicago, Chicago. Regional and national sample surveys and panel studies of politics, education, health and welfare, economics and business, occupations and professions, and community problems.

Public Opinion Survey Unit. University of Missouri, Columbia. Primarily data on Missouri. Election statistics, census data, and sample surveys comprise the archive holdings.

Roper Public Opinion Research Center. Williams College, Williamstown, Mass. Over 7000 sample surveys ranging over a variety of topics—politics, economics, education, and the like. Includes data from American studies as well as data from studies in 43 other countries.

Social Science Data and Program Library Service. University of Wisconsin, Madison. Wisconsin economic data from the 1950s. The holdings also include economic, political, and social data about the 676 incorporated American cities which in 1960 had populations in excess of 25,000.

Survey Research Laboratory. University of Illinois, Urbana. Principal holdings consist of Illinois data on politics, economics, public opinion, and social behavior.

UCLA Political Behavior Archives. University of California, Los Angeles. Special emphasis on Southern California, Austria, Germany, India, and several African states. Data on politics, elite behavior, elections, public opinion, campaign activity, state legislators, and legislative roll calls represent the major collections of the archive.

PART THREE: Scholarly literature (mostly from social sciences with special emphasis upon writings of the 1960s)

A. *METHODOLOGY*

1. Historiography

"10000101010 and All That," *Times Literary Supplement,* (September 8, 1966), 818.

ANDERSON, WALFRED A., *Bibliography of Researches in Rural Sociology.* Ithaca, N.Y.: 1957.

ANDRAE, CARL G., "Population Movements in Sweden: Report on Mass-Data Research Project," *Social Science Information,* 8 (1969), pp. 65–75.

AYDELOTTE, WILLIAM O., "Quantification in History," *American Historical Review,* 71 (1966), pp. 803–825.

BEHRE, OTTO, *Geschichte der Statistik in Brandenburg-Preussen bis zur Gründung des Königlichen Statistischen Bureaus.* Berlin: 1905. Classic history; many excerpts from tables and mss.

CLUBB, JEROME and HOWARD W. ALLEN, "Computers and Historical Studies," *Journal of American History,* 54 (1967), pp. 599–607.

DIÉGUES, M. and B. WOOD, eds., *Social Science in Latin America.* New York: 1967.

DORFMAN, JOSEPH, *The Economic Mind in American Civilization,* 5 vols. New York: 1946–1959.

FUNKHOUSE, H. G., "Historical Development of Graphical Representation of Statistical Data," *OSIRIS,* 3 (1938), pp. 269–404.

GOSNELL, HAROLD F., "Statisticians and Political Scientists," *American Political Science Review,* 27 (1933), pp. 392–409. Good review of research.

GRANTHAM, DEWEY W., "Regional Imagination: Social Scientists and the American South," *Journal of Southern History,* 34 (1968) pp. 2–32.

HARRIS, MARVIN, *The Rise of Anthropological Theory.* New York: 1968. Brief treatment of quantification.

HAYS, SAMUEL P., "Archival Sources for American Political History," *American Archivist,* 28 (1965), pp. 17–30.

HAYS, SAMUEL P., "Computers and Historical Research," *Computers in Humanistic Research*, Edmund A. Bowles, ed. Englewood Cliffs, N.J.: 1967, pp. 62–72.

Instituto Luigi Sturzo, *La Sociologica contemporanea*. Rome: 1968. Important (mostly in English).

KAISER, E., "Von der politischen Arithmetik zur Wirtschaft-und Sozial mathematik," *Schweizerische Zeitschrift fur Volkswirtschaft und Statistik*, 103 (1967), pp. 342–352.

KUHN, THOMAS S., *The Structure of Scientific Revolutions*. Chicago: 1962.

LAQUER, WALTER and GEORGE MOSSE, eds., *The New History*. New York: 1967. See especially J. H. Hexter, "Some American Observations," and Allan Bogue, "United States: the 'New' Political History," for discussions of quantification in recent historiography.

LAVALLE, PLACIDO, et al., "Certain Aspects of the Expansion of Quantitative Methodology in American Geography," *Annals of the Association of American Geographers*, 57 (1967), pp. 423–436.

NIXON, J. W., *A History of the International Statistical Institute: 1885–1960*. The Hague: 1960. See also the *Jubilee Volume of the Statistical Society*, that is, the Royal Statistics Society. London: 1885. On nineteenth century statistical developments.

OBERSCHALL, A. R., *Empirical Social Research in Germany, 1848–1914*. Paris: 1965.

RENOUVIN, PIERRE, "Research in Modern and Contemporary History: Present Trends in France," *Journal of Modern History*, 38 (1966), pp. 1–12. Stresses quantitative studies.

ROWNEY, DON KARL and JAMES Q. GRAHAM, eds., *Quantitative History*. Homewood, Ill.: 1969. Part I.

SAVETH, EDWARD, ed., *American History and the Social Sciences*. New York: 1964. Old but still useful.

SHERIF, MUZAFER and CAROLYN SHERIF, eds., *Interdisciplinary Relationships in the Social Sciences*. Chicago: 1969. See especially essays on History and Sociology by Sidney Aronson and Daniel Calhoun.

SOMIT, ALBERT and JOSEPH TANEHAUS, *The Development of American Political Science*. Boston: 1967.

SWIERENGA, ROBERT P., ed., *Quantification in American History*. New York: 1970. Good anthology, excellent historiographic introductions.

TRUESDALL, LEON E., *The Development of Punch Card Tabulation in the Bureau of the Census: 1890–1940*. Washington: 1965.

WRIGHT, JOHN K., *Human Nature in Geography*. Cambridge: 1966. Delightful essays; see especially those on Gilman and Semple.

WRIGHT, CARROLL D. and W. C. HUNT, *The History and Growth of the United States Census*. Washington: 1900.

2. Research Design

ALKER, HAYWARD R., "The Comparison of Aggregate Political and Social Data: Potentialities and Problems," *Social Sciences Information*, 5 (1966), pp. 63–80.

ALKER, HAYWARD R., "Statistics and Politics: The Need for Causal Data Analysis," in Seymour M. Lipset, ed., *Politics and the Social Sciences*. New York: 1969, pp. 244–313. Review of multivariate statistical aspects of data analysis.

ANDRERLE, O., "A Plea for Theoretical History," *History and Theory*, 4 (1964), pp. 27–56.

BAILEY, NORMAN T. J., *The Mathematical Approach to Biology and Medicine*. New York: 1967. Part One is an excellent discussion of computerized research.

BERND, JOSEPH, ed., *Mathematical Applications in Political Science*, 4 vols. Dallas and Charlottesville, 1965–1969. Important essays on mathematical models.

BERKHOFER, ROBERT F., *A Behavioral Approach to Historical Analysis*. New York: 1969. Important.

CHANMAN, W. and BOSKOFF, A., eds., *Sociology and History*. New York: 1964.

CLARKE, DAVID, *Analytical Archaeology*. London: 1968. Suggestive on taxonomy, time "diachronic" series, and models.

DEMING, WILLIAM E., *Some Theory of Sampling*. New York: 1950. Practical theory; suggestive for actual procedures, especially population surveys.

DOGAN, MATTEI and STEIN ROKKAN, eds., *Quantitative Ecological Analysis*. Cambridge: 1969.

DOLLAR, CHARLES M., "Innovation in Historical Research: A Computer Approach," *Computers and the Humanities*, 3 (1969), pp. 139–152.

DUVERGER, MAURICE, *Introduction to the Social Sciences*, 1st ed. 1959. New York: 1964. French style political science. Good but dated.

FESTINGER, LEON and DANIEL KATZ, eds., *Research Methods in the Behavioral Science*. New York: 1953. Very important, oriented to social psychology.

GALTUNG, JOHAN, *Theory and Methods of Social Research*. New York: 1966. Excellent.

GREER, SCOTT, *The Logic of Social Inquiry*. Chicago: 1968. Ethics and epistemology of social science.

GROSS, LLEWELLYN, ed., *Sociological Theory: Inquiries and Paradigms*. New York: 1967. Good essays on taxonomy, causation, and sociology of knowledge, among other items.

HIRSCHI, TRAVIS and HANAN SELVIN, *Delinquency Research: An Appraisal of Analytic Methods*. New York: 1967. Very helpful.

HOFSTADTER, RICHARD and SEYMOUR M. LIPSET, eds., *Sociology and History: Methods.* New York: 1968. Valuable anthology for quantitative historiography.

HOLT, ROBERT, and JOHN TURNER, *The Methodology of Comparative Research.* New York: 1970.

HYMAN, HERBERT, M., *Survey Design and Analysis.* Glencoe, Ill.: 1955. Good ideas on analysis of dichotomous data.

JAMES, PRESTON EVERETT, "On the Origin and Persistence of Error in Geography," *Association of the American Geographical Annual,* 57 (1967), pp. 1–24.

JAULIN, B., ed., *Calcul et Formalisation dans les sciences de l'homme.* Paris: 1969.

KAPLAN, ABRAHAM, *The Conduct of Inquiry: Methodology for Behavioral Science.* San Francisco: 1964. By far the best book on the abstract theory of methodology.

KRUSKAL, WILLIAM, ed., Mathematical Sciences and Social Science. Englewood Cliffs, N.J.: 1970. Methods and applications; advanced.

LAZARSFELD, PAUL and ROSENBERG, M., eds., *The Language of Social Research.* Glencoe, Ill.: 1955. An excellent anthology.

LAZARSFELD, P. F. and HENRY, N. W., eds., *Reading in Mathematical Social Science.* Chicago: 1966. Important essays on measurement and models.

LERNER, DANIEL and HAROLD D. LASSWELL, eds., *The Policy Sciences; Recent Developments in Scope and Method.* Stanford, Calif.: 1951.

LINDZEY, GARDNER, and ELLIOTT ARONSON, eds., *The Handbook of Social Psychology,* 2nd ed., 5 vols. Reading, Mass.: 1968. Contents: vol. 1: history, theory; vol. 2: research methods (including data analysis, simulation, content analysis); vol. 3: the individual (for example, attitudes, socialization); vol. 4: group phenomena (small groups, leadership, social structure, collective psychology, national character); vol. 5: applied social psychology (for example, mass media, economics, political behavior, international relations, religion, prejudice). Thorough review of discipline.

LIPSET, SEYMOUR M., ed., *Politics and the Social Sciences.* New York: 1969. Interdisciplinary relations; two essays by Richard Jensen consider the growth of quantitative methods in history and political science.

MCKINNEY, JOHN C., *Constructive Typology and Social Theory.* New York: 1965. Valuable.

MELITZ, J., "Friedman and Machlup on the Significance of Testing Economic Assumption," *Journal of Political Economy,* 73 (1965), pp. 37–60.

MERRITT, RICHARD, and STEIN ROKKAN, eds., *Comparing Nations: The Use of Quantitative Data in Cross-National Research*. New Haven, Conn.: 1966. Important collection of articles, especially those by Scheuch on ecological fallacy, and Alker and Russett on inequality.

MURPHY, GEORGE G. S., "Historical Investigation and Automatic Data Processing Equipment," *Computers and the Humanities*, 3 (1968), pp. 1–13.

NAGEL, ERNEST, *The Structure of Science*. New York: 1961. Important, relates social sciences and history to natural sciences.

OGBURN, W. F., *On Culture and Social Change: Selected Papers*, edited and with an introduction by O. D. Duncan. Chicago: 1964. Important papers of leading statistical sociologist with historical bent.

ORCUTT, GUY, HAROLD WATTS, and JOHN EDWARDS, "Data Aggregation and Information Loss," *American Economic Review*, 58 (1968), pp. 773–787, Simulation study of the ecological fallacy.

PEARSON, KARL, *The Grammar of Science* (1st ed., 1892). Very influential but extreme statement in favor of scientific method in all research.

PHILLIPS, B. S., *Social Research: Strategy and Tactics*. New York: 1966. Excellent.

PRZEWORSKI, ADAM, and HENRY TEUNE, *The Logic of Comparative Social Inquiry*. New York: 1969.

ROSENBERG, MORRIS, *The Logic of Survey Analysis*. New York: 1968. Valuable on multivariate analysis of dichotomous data.

RUDNER, RICHARD, *Philosophy of Social Science*. Englewood Cliffs, N.J.: 1966. Rather abstract.

SELVIN, HANAN C., "A Critique of Tests of Significance," *American Sociological Review*, 22 (1957), pp. 519–527.

SIMON, HERBERT A., *Models of Man, Social and Rational; Mathematical Essays on Rational Human Behavior in a Social Setting*. New York: 1961. Important, especially causal models.

SJOBERG, GIDEON and ROGER NETT, *A Methodology for Social Science*. New York: 1969.

Social Science Research Council, Bulletin 64, *The Social Sciences in Historical Study*. New York: 1954. Very influential but increasingly outdated; good statement of historical method.

STOUFFER, SAMUEL, *Social Research to Test Ideas*. Glencoe, Ill.: 1962. Collected essays of foremost statistical sociologist.

TUKEY, JOHN W., "The Future of Data Analysis," *Annals of Mathematical Statistics*, 33 (1962), pp. 1–67. Advanced.

3. Statistics

ALKER, HAYWARD R., JR., *Mathematics and Politics.* New York: 1965. Important but too condensed.

ARKIN, HERBERT, and RAYMOND COLTON, *Tables for Statisticians.* New York: 1963. College Outline Series. Very useful.

BARBAT, M., *Mathematiques des sciences humaines,* 2 vols. Paris: 1967.

BARTHOLOMEW, DAVID J., *Stochastic Models for Social Process.* New York: 1967.

BARTOS, OTOMAR J., *Simple Models of Group Behavior.* New York: 1967. Markov chains, game theory, and so forth. Valuable. Little math required.

BLALOCK, HUBERT M., "Causal Inferences, Closed Populations and Measures of Association," *American Political Science Review.* 61 (1967), pp. 130–136.

BLALOCK, HUBERT, and A. BLALOCK, eds., *Methodology in Social Research.* New York: 1968. Very important; essays on measurement. causation, error, social status, sampling, and model building.

BLALOCK, HUBERT M., "Some Implications of Random Measurement Error for Causal Inferences," *American Journal of Sociology,* 71 (1965), pp. 37–47; "Theory Building and the Statistical Concept of Interaction," *American Sociological Review,* 30 (1965), pp. 374–380.

BOUDON, RAYMOND, "Method of Linear Causal Analysis: Dependence Analysis," *American Sociological Review,* 30 (1965), pp. 365–374.

BOYLE, R. P., "Causal Theory and Statistical Measures of Effect: A Convergence," *American Sociological Review,* 31 (1966), pp. 843–851.

CATTELL, R. B., ed., *Handbook of Multivariate Experimental Psychology.* Chicago: 1965. Important, massive summary of Cattell approach, emphasizing theoretical foundations of factor analysis.

CHADWICK, RICHARD W., and K. W. DEUTSCH, "Doubling Time and Half Life: Two Suggested Conventions for Describing Rates of Change in Social Science Data," *American Behavioral Scientist,* 11 (1968), N53–N511. Useful.

COLEMAN, JAMES S., *Introduction to Mathematical Sociology.* New York: 1964. Disappointing and difficult.

CONWAY, FREDA, *Sampling.* London: 1967. Introductory text; useful.

COOLEY, W. W., and P. R. LOHNES, *Multivariate Procedures for the Behavioral Sciences.* New York: 1962. Computer oriented; difficult for one not familiar with rudiments of matrix algebra.

COSTNER, HERBERT LEE, "Criteria for Measures of Association," *American Sociological Review,* 30 (1965), pp. 341–53. Very important; introduces concept of proportional reduction of error.

DAVIS, HAROLD T., *Political Statistics.* Evanston, Ill.: 1954.

DUNCAN, OTIS D. "Path Analysis: Sociological Examples," *American Journal of Sociology,* 72 (1966), pp. 1–16.

FENNESSEY, JAMES, "The General Linear Model," *American Journal of Sociology,* 74 (1968), pp. 1–27. Excellent review of the varieties of regression analysis.

FISCHER, FRANKLIN M., *The Identification Problem in Econometrics.* New York: 1966. Advanced.

FLAMENT, CLAUDE, *Theorie des graphes et structures sociales.* Paris: 1965.

FOX, KARL S., *Intermediate Economic Statistics.* New York: 1968. Good introduction to econometrics.

FREUND, JOHN E., *Modern Elementary Statistics,* 2nd ed. Englewood Cliffs, N.J.: 1966. Theoretical text, not cookbook.

GOLD, DAVID. "Statistical Tests and Substantive Significance," *American Sociologist,* 4 (1969), pp. 42–46.

GOODMAN, LEO, "Some Alternatives to Ecological Correlation," *American Journal of Sociology,* 64 (1959), pp. 610–625. Proves that in general slope of ecological regression line (b) equals individual correlation coefficient (ϕ).

GOODMAN, LEO and WILLIAM KRUSKAL, "Measures of Association for Cross Classifications," *Journal of the American Statistical Association.* 49 (1954), pp. 732–764. Classic.

HAYS, WILLIAM, *Statistics for Psychologists.* New York: 1963. Very useful for theoretical grounds (set theory, probability); not too good a cookbook.

HOLLAND, JANET and M.D. STEUER, *Mathematical Sociology: A Selective Annotated Bibliography.* London: 1969. 451 items.

HORST, P., *Matrix Algebra for Social Scientists.* 1963. Good introduction.

JOHNSTON, J., *Econometric Methods.* New York: 1963. Good, but theoretical.

KEMENY, JOHN G., et al., *Introduction to Finite Mathematics.* Englewood Cliffs, N.J.: 1966. Best introduction to necessary math.

KENDALL, MAURICE and ALLAN STUART, *The Advanced Theory of Statistics,* 3 vols. London, New York: 1967. Chapters on time series (in vol. 3) are readable. The rest of the series is very advanced.

KENDALL, MAURICE GEORGE, and WILLIAM R. BUCKLAND, *A Dictionary of Statistical Terms,* 2nd ed. New York: 1960. With English-

French-German-Italian dictionary; handy when confusion over terminology arises; normative guide to usage, but few formulas.

KEY, V. O., JR., *A Primer of Statistics for Political Scientists.* New York: 1954. Useful only for time series analysis of aggregate election returns.

MCGINNIS, ROBERT, *Mathematical Foundations for Social Analysis.* Indianapolis: 1965. Good starter on matrix algebra.

MALINVAUD, E., *Statistical Methods of Econometrics.* Chicago: 1966. Advanced.

MORRISON, DONALD F., *Multivariate Statistical Methods.* New York: 1967. Good, but not easy.

QUENOVILLE, M., *Rapid Statistical Calculations.* New York: 1960. Simple and useful.

SCHMID, CALVIN, *Handbook of Graphic Presentation.* New York: 1954. Very useful for designing presentations.

SEIGLE, DAVID, "Some Aids in the Handling of Missing Data," *Social Science Information,* 6 (1967), pp. 133–150. Helpful.

SIEGEL, SIDNEY, *Non-Parametric Statistics for the Behavioral Sciences.* New York: 1956. Convenient compilation, but not often useful for historians.

SOMERS, ROBERT, "An Approach to the Multivariate Analysis of Ordinal Data," *American Sociological Review,* 33 (1968), pp. 971–977.

THEIL, HENRI, *Economic and Information Theory.* Chicago: 1967. Most useful for new, decomposable measure of inequality.

THOMAS, EDWIN N., and D. L. ANDERSON, "Additional Comments on Weighting Values in Correlation Analysis of a Real Data," *Association of the American Geographical Annual,* 55 (1965), pp. 482–505.

WINCH, ROBERT and DONALD CAMPBELL, "Proof? No. Evidence? Yes. The Significance of Tests of Significance," *American Sociologist* 4 (1969), pp. 140–143. Tests of significance can be used in non-sampling situations by assuming the possibility of "scrambling" the pattern.

YULE, G. UDNY and MAURICE G. KENDALL, *An Introduction to the Theory of Statistics,* 14th ed. London: 1950. The grand old classic; still good for dichotomous data.

4. Statistics Texts (all are recommended)

BLALOCK, HUBERT, *Social Statistics.* New York: 1960.

CROWDEN, F., D. CROXTON, and M. KLEIN, *Applied General Statistics.* New York: 1967. Good for time series.

DIAMOND, SOLOMON, *Information and Error.* New York: 1959. Psychology; wit; new slants on old ideas.

DIXON, WILFRED, and F. MASSEY, *Introduction to Statistical Analysis.* New York: 1969. Good on shortcuts.

HUFF, D., and I. GEIS, *How to Lie with Statistics.* New York: 1954. Entertaining and useful on fallacies.

MUELLER, JOHN, and KARL SCHUESSLER, *Statistical Reasoning in Sociology.* Boston: 1963. Good for class text.

PALUMBO, DENNIS J., *Statistics in Political and Behavioral Science.* New York: 1969. Excellent. Accompanying workbook is helpful.

WALKER, HELEN and JOSEPH LEV, *Elementary Statistical Methods,* 2nd ed. New York: 1958. One of the best "first" books. *Statistical Inference.* New York: 1953. One of the best of the slightly more advanced texts.

WEISS, ROBERT S., *Statistics in Social Research.* New York: 1968. Fair elementary text.

5. Electronic Data Processing

BAER, MICHAEL A., and HARMON ZEIGLER, "Computers and Political Science: A Review Article," *Computers and Humanities,* (1967), pp. 135–143.

BATES, FRANK, and MARY DOUGLAS, *Programming Language One.* Englewood Cliffs, N.J.: 1967. Survey of PL/1 for IBM 360.

BISCO, RALPH, "Social Science Data Archives: A Review of Developments," *American Political Science Review,* 55 (1966), pp. 93–109.

BORKO, H., ed., *Computer Applications in the Behavioral Sciences.* Englewood Cliffs, N.J.: 1962. Older introduction to nature of computers; interesting chapters on computation, models, simulation in variety of fields (not history).

CASEY, ROBERT S., ed., *Punched Cards: Their Application to Science and Industry,* 2nd ed. New York: 1958. Stresses utility of edge-punched cards and their hand manipulation by needles for small research projects not needing extensive recoding or statistical measures.

DIMITRY, D., and T. MOTT, JR., *Introduction to Fortran IV Programming.* New York: 1966.

HORN, JACK, *Computer and Data Processing Dictionary and Guide.* Englewood Cliffs, N.J.: 1966.

IBM Corporation, *Introduction to IBM DATA-Processing Systems.* (N.D.)

IBM Corporation, *Fortran IV for the IBM System 360: Programmed Instruction Course* (1965).

JANDA, KENNETH, *Data Processing: Applications to Political Research.* Evanston, Ill.: 1969.

MCCRACKEN, D. D., *Guide to Fortran IV Programming.* New York: 1965.

ORGANICK, E. I., *Fortran IV Primer.* Reading, Mass.: 1966.

Programs and Manuals: Dixon, Wilfred, *UCLA Biomedical Programs* (Los Angeles, 1965). Useful statistical routines. J. Sonquist, and J. Morgan, *The Detection of Interaction Effects* (Ann Arbor, 1965), A.I.D. manual. F. Andrews, J. Morgan, and J. Sonquist, *Multiple Classification Analysis,* (Ann Arbor, 1967), MCA manual. Current programs are written up in *Behavioral Science, Educational and Psychological Measurement.* Every computer center has its own manual, and probably the IBM scientific package, too.

SIPPL, CHARLES J., *Computer Dictionary and Handbook.* Indianapolis: 1966. *Computer Dictionary.* Indianapolis: 1966. Very convenient guide, not only to terminology but to hardware, software, various a-plied problems, and nature of computer installations.

SLATER, LUCY JOAN, *FORTRAN Programming for Economists.* Cambridge, England: 1967.

STERLING, THEODOR, and SEYMOUR POLLACK, *Computers and the Life Sciences.* New York: 1965. Important demonstration of computers' potential in searching for and classifying data.

STERLING, THEODOR, and SEYMOUR POLLACK, *Introduction to Statistical Data Processing.* Englewood Cliffs, N.J.: 1968.

STERLING, THEODOR, and SEYMOUR POLLACK, *A Guide to PL/1.* New York: 1969. Best guide for social researchers.

VELDMAN, DONALD J., *FORTRAN Programming for the Behavioral Sciences.* New York: 1967. Very good guide to canned programs.

WEINBERG, GERALD, *PL/1 Programming Primer.* New York: 1966.

6. Geography

ALEXANDER, JOHN, *Economic Geography.* Englewood Cliffs, N.J.: 1963. Very useful text, for it introduces statistical techniques, shows possibilities of mapping, and gives recent theoretical frameworks.

BERRY, BRIAN J., and THOMAS D. HANKINS, *A Bibliographic Guide to the Economic Regions of the U.S.* Chicago: 1965. Very important annotated guide to methodology along with numerous items on specific regions.

BERRY, BRIAN, and DUANE MARBLE, *Spatial Analysis: A Reader in Statistical Geography.* Englewood Cliffs, N.J.: 1968. Very suggestive.

BERRY, BRIAN J., *Geography of Market Centers and Retail Distribution.* Englewood Cliffs, N.J.: 1967.

BOGUE, DONALD JOSEPH and CALVIN L. BEALE. *Economic Areas of the United States.* New York: 1961. Detailed compilation of data.

BROWN, LAWRENCE J., *Diffusion Processes and Location.* Philadelphia: 1968.

CHORLEY, RICHARD and PETER HACKETT, eds., *Models in Geography.* London: 1967. Important anthology, especially first third.

CLARKE, J. I., "Statistical Map-Reading," *Geography,* 44 (1959), pp. 96–104. On optical illusions and misperceptions.

DOHRS, FRED, and LAWRENCE SOMMERS, *Introduction to Geography: Selected Readings; Cultural Geography: Selected Readings.* New York: 1968. Two very readable and suggestive anthologies.

DUNCAN, OTIS D., A. CUZZART, and B. DUNCAN, *Statistical Geography.* New York: 1961. On ecological fallacy.

GARRISON, W. L. and D. F. MARBLE, eds., *Quantitative Geography,* 2 vols. Evanston, Ill.: 1967. Very suggestive.

GINSBURG, NORTON SYDNEY, *Atlas of Economic Development.* Chicago: 1961.

HÄGERSTRAND, TORSTEN, *Innovation Diffusion as a Spatial Process.* Chicago: 1967. Major geographic models.

HAGGETT, PETER, *Locational Analysis in Human Geography.* London: 1965. Important; quantitative approach to location.

HOOVER, EDGAR M., *The Location of Economic Activity.* New York: 1948.

ISARD, W., *Methods of Regional Analysis: An Introduction to Regional Science.* New York: 1960.

KING, LESLIE J., *Statistical Geography.* New York: 1969.

KRALLERT, W., "Methodische Probleme der Völker-und Sprachen-Karten," *International Yearbook of Cartography,* 1. London: 1961, pp. 99–120. On methods of ethnographic mapping.

LÖSCH, A., *The Economics of Location.* New Haven, Conn.: 1954. Suggested classic.

MCDONALD, J. R., "Region: Its Conception, Design and Limitations," *Association of the American Geographical Annual,* 56 (1966), pp. 516–528.

MONKHOUSE, F. J. and H. R. WILKINSON, *Maps and Diagrams.* New York: 1963. Excellent manual.

NEWLING, B. E., "Urban Growth and Spatial Structure: Mathematical Models and Empirical Evidence," *Geographic Review,* 56 (1966), pp. 213–225.

OLSSON, GUNNAR, *Distance and Human Interaction: A Review and Bibliography.* Philadelphia: 1965.

PRED, ALLAN, *Behavior and Location: Foundations for a Geographic and Dynamic Location Theory.* Lund, Sweden: 1967.

PRED, ALLAN R., *The Spatial Dynamics of US Urban-Industrial Growth, 1800–1914: Interpretive and Theoretical Essays.* Cambridge: 1966. Important new statistical models.

RAPOPORT, ANATOL, "Spread of Information through a Population with Socio-Structural Bias," *Bulletin of Mathematical Biophysics,* 15 (1953), pp. 523–540. Based on epidemiological models of diffusion. Suggestive.

SAVAGE, RICHARD and KARL DEUTSCH, "A Statistical Model for the Gross Analysis of Transaction Flows," *Econometrica* **28** (1960, pp. 551–572.)

SOJA, EDWARD, *The Geography of Modernization in Kenya.* Syracuse: 1968. Impressive technique.

YEATES, MAURICE H., *An Introduction to Quantitative Analysis in Economic Geography.* New York: 1968.

7. Content Analysis and Information Retrieval

BERELSON, BERNARD, *Content Analysis in Communication Research.* Glencoe, Ill.: 1952.

BISCO, R. L. "Social Science Data Archives: A Review of Developments," *American Political Science Review,* 59 (1965), pp. 93–109.

BUDD, R. W., R. K. THORPE, and L. DONOHEW, *Content Analysis of Communications.* New York: 1967. Good for journalism research.

BOURNE, CHARLES P., *Methods of Information Handling.* New York: 1963.

GARRATY, JOHN A., "The Application of Content Analysis to Biography and History," in Ithiel de Sola Pool, ed., *Trends in Content Analysis.* Urbana, Ill.: 1959, pp. 171–187.

GERBNER, GEORGE, et al. eds., *The Analysis of Communication Content.* New York: 1969. Summary of research.

HOLSTI, OLE R., *Content Analysis for the Social Sciences and Humanities.* Reading, Mass.: 1969. Good survey.

HOLSTI, OLE R., RICHARD A. BRODY, and ROBERT C. NORTH, "Affect and Action in International Reaction Models: Empirical Materials from the 1962 Cuban Crisis," *Journal of Peace Research* (1964), pp. 170–190.

HOLSTI, O. R., "The 1914 Case," *American Political Science Review,* 59 (1965), pp. 365 378.

HOLSTI, OLE R., "Content Analysis," in Gardner Lindzey and Elliot Aronson, eds., *The Handbook of Social Psychology,* vol. 2 Reading, Mass.: 1968.

JANDA, KENNETH, *Information Retrieval.* Indianapolis: 1968. Basic introduction to KWIC, SDI, and other systems.

JANSEN, B. DOUGLASS, "A System for Content Analysis by Computer of International Communications for Selected Categories of Action," *American Behavioral Scientist,* 9 (1966), pp. 28–32.

KENT, ALLEN and HAROLD LANCOUR, eds., *Encyclopedia of Library and Information Science.* New York: 1968. In progress. A multivolume compilation with superb coverage.

LANTZ, HERMAN, et al., "Pre-Industrial Patterns in the Colonial Family in America: A Content Analysis of Colonial Magazines," *American Sociological Review*, 33 (1968), pp. 413–426.

MERRITT, RICHARD L., "The Emergence of American Nationalism: A Quantitative Approach," *American Quarterly*, 17 (1965), pp. 319–371. Content analysis of five colonial newspapers expanded in *Symbols of an American Community*. New Haven, Conn.: 1966.

MERRITT, RICHARD L., "The Representational Model in Content Analysis," in Joseph L. Bernd, ed., *Mathematical Applications in Political Science*, II. Dallas: 1966, pp. 44–71.

MERRITT, RICHARD, and STEIN ROKKAN, eds., *Data Archives for the Social Sciences*. Paris: 1966.

NORTH, ROBERT C., OLE R. HOLSTI, M. GEORGE ZANINOVICH, and DINA A. ZINNES, *Content Analysis: A Handbook with Application to the Study of International Crisis*. Evanston, Ill.: 1963.

PERMAN, DAGMAR H., ed., *Bibliography and the Historian: The Conference at Belmont*. Santa Barbara, Calif.: 1968.

POOL, ITHIEL DE SOLA, ed., *Trends in Content Analysis*. Urbana, Ill.: 1959.

ROKKAN, STEIN and FRANK AAREBROT, "The Norwegian Archive of Historical Ecological Data," *Social Science Information*, 8 (1969), pp. 77–84. Important for technique.

SANDERS, L., "A Content Analysis of President Kennedy's First Six Press Conferences," *Journal Quarterly*, 42 (1965), pp. 114–116.

SHARP, H., ed., *Readings in Information Retrieval*. New York: 1965.

SHARP, J. R., *Some Fundamentals of Information Retrieval*. New York, London: 1965.

STONE, PHILIP J., et al., *The General Inquirer: A Computer Approach to Content Analysis*. Cambridge: 1966. Important, computerized methods.

8. Clustering Methods

BALL, G. H., "Data Analysis in the Social Sciences," *Proceedings— Fall Joint Computer Conference*, (1965).

BERRY, B. J. L., "A Method for Deriving Multifactor Uniform Regions," *Przeglad Geograficzny*, 33 (1961), pp. 263–282.

CAMIN, JOSEPH, and ROBERT SOKAL, "A Method for Deducing Branching Sequences in Phylogeny," *Evolution*, 19 (1965), pp. 311–326. Important; uses dendrograms to suggest evolutionary sequences (for organisms that, of course, evolve in suitable biological fashion). Application to history suggestive, but dangerous.

CATTELL, RAYMOND, and MALCOLM COULTER, "Principle of Behavioural Taxonomy and the Mathematical Basis of the Taxonome

Computer Program," *British Journal of Mathematical and Statistical Psychology,* 19 (1966), pp. 237–269. Cluster analysis based on similarity of profiles.

HARMAN, HARRY H., *Modern Factor Analysis,* 2nd ed. Chicago: 1967. Difficult; easier is Rudolph Rummell, *Applied Factor Analysis.* Evanston, Ill.: 1970.

HARTIGAN, J. A., "Representation of Similarity Matrices by Trees," *Journal of American Statistical Association,* 62 (1967), pp. 1140–1158. Dendrograms, using clusters of states based on election returns as example.

HORST, PAUL, *Factor Analysis of Data Matrices.* New York: 1965. Advanced computer programs.

KING, BENJAMIN, "Step-Wise Clustering Procedures," *Journal of the American Statistical Association,* 72 (1967), pp. 86–101.

JONES, KENNETH J., "Problems of Grouping Individuals and the Method of Modality," *Behavioral Science,* 13 (1968), pp. 496–511. Review of clustering methods and computer programs.

MAYR, ERNST, "Numerical Phenetics and Taxonomic Theory," *Systematic Zoology,* 14 (1965), pp. 73–97.

MCMURRAY, C. D., "Choice of Correlation Coefficient and Its Effect on Factor Structure in the Analysis of Dichotomous Political Data," *Research Reports in Social Science,* 9 (1966), pp. 38–54.

MCQUITTY, LOUIS L., "Capabilities and Improvements of Linkage Analysis as a Clustering Method," *Educational & Psychological Measurement,* 28 (1968), pp. 211–238.

MCQUITTY, LOUIS L., "Elementary Factor Analysis," *Psychological Reports,* 9 (1961), pp. 71–78.

MCQUITTY, LOUIS L., "A Mutual Development of Some Typological Theories and Patterns-Analytic Methods," *Educational & Psychological Measurement,* 27, pp. 21–46.

MEGEE, MARY, "On Economic Growth and the Factor Analysis Method," *Southern Economic Journal,* 31 (1965), pp. 215–228. "R" factor analysis.

MORGAN, JAMES and HON SONQUIST, *Multiple Classification Analysis.* Ann Arbor, Mich.: 1967. Important computer programs.

MURPHEY, GEORGE G., and M. G. MUELLER, "On Making Historical Techniques More Specific: Real Types Constructed with a Computer," *History & Theory,* 6 (1967), pp. 14–32. A kind of multivariate classification algorithm.

NUNNALLY, J. and P. H. A. SNEATH, "The Analysis of Profile Data," *Psychological Bulletin,* 59 (1962), pp. 311–319.

SHEPARD, ROGER and DOUGLAS CARROLL, "Parametric Representation of Nonlinear Data Structures," in P. R. Krishnaiah, ed., *Multivariate Analysis.* New York: 1966. Pp. 561–592. Readable introduction to new and very important Shepard and Kruskal family of techniques.

SOKAL, ROBERT, and P. H. A. SNEATH, *Principles of Numerical Taxonomy.* San Francisco: 1963. Important manifesto (in biology), with suggestive applications; see also Robert Sokal, "Numerical Taxonomy," in *Scientific American,* 215 (1966), pp. 106–116.

TRYON, R. C., and D. E. BAILEY, "The BC Try Computer System of Cluster and Factor Analysis," *Multivariate Behavioral Research,* 1 (1966), pp. 95–111.

WARD, J. H., JR., "Hierarchical Grouping to Optimize an Objective Function," *Journal of The American Statistical Association,* 58 (1963), pp. 236–244. Classic in clustering theory.

B. *POLITICAL STUDIES*

1. Methods

BELL, RODERICK, et al., eds., *Political Power: A Reader in Theory and Research.* New York: 1969.

BOULDING, KENNETH, *Conflict and Defense.* New York: 1962.

CHARLESWORTH, JAMES, ed., *Contemporary Political Analysis.* New York: 1967.

DAHL, ROBERT A., *Modern Political Analysis.* Englewood Cliffs, N.J.: 1970.

DAVIS, OTTO A. and others, "Theory of the Budgetary Process," *American Political Science Review,* 60 (1966), 529–547; 61 (1967), pp. 152–153. Important model for Congress.

DEUTSCH, KARL W., "Social Mobilization and Political Development," *American Political Science Review,* 55 (1961), pp. 493–515.

DEUTSCH, KARL W., *The Nerves of Government.* New York: 1966.

DOWNS, ANTHONY, *An Economic Theory of Democracy.* New York: 1957.

EULAU, HEINZ, et al., *Political Behavior: A Reader in Theory and Research.* Glencoe, Ill.: 1956.

GARCEAU, OLIVER, ed., *Political Research and Political Theory.* Cambridge: 1968.

GARVEY, G., "Theory of Party Equilibrium," *American Political Science Review,* 60 (1966), pp. 29–38. Deepening of Downsan models.

GOLEMBIEWSKI, ROBERT T., WILLIAM WELSH, and WILLIAM CROTTY, *A Methodological Primer for Political Scientists.* Chicago: 1969. Suggestive, especially for study of administration, but weak on quantification.

HAYS, SAMUEL P., "Social Analysis of American Political History, 1880–1920," *Political Science Quarterly,* 80 (1965), pp. 373–394.

KLINGBERG, F. L. "Predicting the Termination of War: Battle, Casualties and Population Losses," *Journal of Conflict Resolution,* 10 (1966), pp. 129–171.

MACKENZIE, W. J. M., *Politics and Social Science.* Baltimore: 1967.

MONSMA, STEPHEN V., and JACK R. VAN DER SLIK, *American Politics: Research and Readings.* New York: 1970.

OZOUF, JACQUES, "Mesure et Demesure: L'Etude de l'Opinion," *Annales,* 21 (1966), pp. 342–345. On measuring historic public opinion.

PFEIFFER, DAVID G., "Measurement of Inter-party Competition and Systemic Stability," *American Political Science Review,* 61 (1967), pp. 457–467.

RETZLAFF, RALPH H., "Use of Aggregate Data in Comparative Political Analysis," *Journal of Politics,* 27 (1965), pp. 797–814.

RICE, STUART A., *Quantitative Methods in Politics.* New York: 1928. Classic.

SCHUBERT, GLENDON, ed., *Judicial Behavior.* New York: 1965. Superb anthology of techniques for roll call analysis and collector biography.

ULMER, S. SIDNEY, ed., *Introductory Readings in Political Behavior.* Chicago: 1961. Excellent anthology.

WISEMAN, H., *Political Systems: Some Sociological Approaches.* New York: 1966. Useful review of current theories.

2. Comparative and International Political Science

BANKS, ARTHUR S., and ROBERT B. TEXTOR, *A Cross-Polity Survey.* Cambridge: 1963. Mostly computer printout and cross-tabulations; fascinating stuff.

CRITTENDEN, JOHN, "Dimensions of Modernization in the American States," *American Political Science Review,* 61 (1967), pp. 989–1001.

CUTRIGHT, PHILLIPS, "National Political Development: Measurement and Analysis," *American Sociological Review,* 28 (1963), pp. 253–264.

DAHL, ROBERT A., ed., *Political Oppositions in Western Democracies.* New Haven, Conn.: 1965.

DEUTSCH, KARL W., *Introduction to International Relations.* Englewood Cliffs, N.J.: 1968.

DEUTSCH, KARL W., "Towards an Inventory of Basic Trends and Patterns in Comparative and International Politics," *American Political Science Review,* 54 (1960), pp. 34–57.

FEIERABEND, I. K., and R. L. FEIERABEND, "Aggressive Behavior within Polities, 1948–1962: A Cross-National Study," *Journal of Conflict Resolution,* 10 (1966), pp. 249–271. Factor analysis.

FITZGIBBON, RUSSEL H., "Measuring Democratic Change in Latin America," *Journal of Politics*, 29 (1967), pp. 129–166.

HOLSTI, K. J., *International Politics: A Framework for Analysis*. Englewood Cliffs, N.J.: 1967.

LINZ, JUAN J., "The Party System of Spain: Past and Future," in Seymour M. Lipset and Stein Rokkan, *Party Systems and Voter Alignments: Cross National Perspectives*. New York: 1967.

MCCRONE, DONALD J., and C. F. CNUDDE, "Toward a Communications Theory of Democratic Political Development: A Causal Model," *American Political Science Review*, 61 (1967), pp. 72–79.

MCDONALD, RONALD H., "Electoral Systems, Party Representation, and Political Change in Latin America," *Western Political Quarterly*, 20 (1967), pp. 694–708.

MERRITT, RICHARD L., and STEIN ROKKAN, ed., *Comparing Nations: The Use of Quantitative Data in Cross-National Research*. New Haven: 1966.

MERRITT, RICHARD L., *Western European Perspectives on International Affairs: Public Opinion Studies and Evaluations*. New York: 1967.

NILSON, STEN SPARRE, "Measurement and Models in the Study of Stability," *World Politics*, 20 (1967), pp. 1–29.

PATTERSON, SAMUEL CHARLES, "Political Cultures of the American States," *Journal of Politics*, 30 (1968), pp. 187–209.

PUTNAM, ROBERT K., "Toward Explaining Military Intervention in Latin American Politics," *World Politics*, 20 (1967), pp. 83–110.

ROKKAN, STEIN, ed., *Comparative Research Across Cultures and Nations*. Paris: 1968.

RUMMEL, R. J., "Dimensions of Conflict Behavior within Nations, 1946–59," *Journal of Conflict Resolution*, 10 (1966), pp. 65–73. Factor analysis.

RUMMEL, R. J., "Dimensions of Dyadic War, 1820–1952," *Journal of Conflict Resolution*, 11 (1967), pp. 176–183.

RUMMEL, R., J. SAWYER, R. TANTER, and H. GUETZKOW, *Dimensions of Nations* (1969). Factor analysis.

RUSSETT, BRUCE M., ed., *Economic Theories of International Politics*. Chicago: 1969.

SINGER, JOEL DAVID, and MELVIN SMALL, "Composition and Status Ordering of the International System: 1815–1940," *World Politics*, 18 (1966), pp. 236–282.

SINGER, J. DAVID, ed., *Quantitative International Politics*. New York: 1968. Major group of articles, with historical slant.

TANTER, RAYMOND, "Dimensions of Conflict Behavior Within and Between Nations, 1958–1960," *Journal of Conflict Resolution*, 10 (1966), pp. 41–64. Multivariate analysis.

TANTER, RAYMOND, and MANUS MIDLARSKY, "Theory of Revolution," *Journal of Conflict Resolution*, 11 (1967), pp. 264–280. Causal models.

TAYLOR, CHARLES, ed., *Aggregate Data Analysis: Political and Social Indicator in Cross-National Research*. Paris: 1968.

TEUNE, H., and S. SYNNESTVEDT, "Measuring International Alignments," University of Pennsylvania Foreign Policy Research Institute (Philadelphia, 1965).

WRIGHT, QUINCY, *A Study of War*, 2 vols. Chicago: 1942; 2nd ed., 1965. Basic.

3. Elections and Parties

ALEXANDER, THOMAS B., et al., "The Basis of Alabama's Ante-Bellum Two Party System: A Case Study in Party Alignment and Voter Response in the Traditional Two-Party System of the U.S. by Quantitative Analysis Methods," *Alabama Review*, 19 (1966), pp. 243–276.

ALLARDT, ERIK, and STEIN ROKKAN, ed., *Mass Politics*. New York: 1970. European political sociology.

BAGGALEY, ANDREW, "Religious Influence on Wisconsin Voting, 1928–1960," *American Political Science Review*, 56 (1962), pp. 66–77. Ecological.

BAXTER, CRAIG, *District Voting Trends in India*. New York: 1970. Covers 1951–1969.

BENSON, LEE, *The Concept of Jacksonian Democracy: New York as a Test Case*. New York: 1961. Statistically weak, and too heavily under the spell of Lazarsfeld.

BENSON, LEE, "Research Problems in American Political Historiography," in Mirra Komarovsky, ed., *Common Frontiers in the Social Sciences*, Glencoe, Ill.: 1957, pp. 113–183. Poses problems of dating turning points and locating events.

BURNHAM, WALTER DEAN, "Changing Shape of the American Political Universe," *American Political Science Review*, 59 (1965), pp. 7–28. Extremely important.

———, *Critical Elections and the Future of American Politics*. New York: 1971. Wide-ranging.

BUTLER, DAVID E., and DONALD E. STOKES, *Political Change in Britain*. New York: 1969. Based primarily on survey data, with attention to change over generations.

CAMPBELL, ANGUS, PHILIP CONVERSE, WARREN MILLER, and DONALD STOKES, *Elections and the Political Order*. New York: 1966. Essays, some of them historical; very important, but not quite as much as the same authors' *The American Voter*, 1960.

CHAMBERS, WILLIAM, and W. DEAN BURNHAM, eds., *American Party Systems*. New York: 1967.

CLEM, A., *West River Voting Patterns.* Pierre, 1965. On recent South Dakota elections.

CLUBB, JEROME, and HOWARD W. ALLEN, "The Cities and the Election of 1928," *American Historical Review,* 74 (1969), pp. 1205–1220.

COSMAN, BERNARD, "Religion and Race in Louisiana Presidential Politics, 1960," *Southwestern Social Science Quarterly,* 43 (1962–63), pp. 235–241.

CRITTENDEN, JOHN, "Aging and Party Affiliation," *Public Opinion Quarterly,* 26 (1962), pp. 648–657. Historical use of Gallup polls.

CUMMINGS, MILTON C., *Congressmen and the Electorate.* New York: 1966.

CUMMINGS, MILTON C., ed., *The National Election of 1964,* Washington: 1966. Numerous statistical studies.

DANDO, W. A., "Map of the Election to the Russian Constituent Assembly," *Slavic Review,* 25 (1966), pp. 314–319.

DANIELS, GEORGE H., "Immigrant Vote in the 1860 Election: The Case of Iowa," *Mid-America,* 44 (1962), pp. 159–162.

DUNBABIN, J. P. D., "Parliamentary Elections in Great Britain, 1868–1900: A Psephological Note," *English Historical Review,* 81 (1966), pp. 82–99.

DUVERGER, MAURICE, *Political Parties.* New York: 1954.

DYKSTRA, ROBERT R., and HARLAN HAHN, "Northern Voters and Negro Suffrage: The Case of Iowa, 1868," *Public Opinion Quarterly,* 32 (1968), pp. 202–215.

EDGEWORTH, FRANK W., "Miscellaneous Applications of the Calculus of Probabilities," *Journal of the Royal Statistical Society,* 61 (1898), pp. 534–544. Pioneering election analysis of aggregate data. Did not attempt to correlate with social data (that was Turner's innovation a few years before).

ELDERSVELD, SAMUEL J., *Political Affiliation in Metropolitan Detroit.* Ann Arbor, Mich.: 1957.

EPSTEIN, LEON D., *Politics in Wisconsin.* Madison, Wisc.: 1958.

ERSKINE, H. G., "Polls: Organized Religion," *Public Opinion Quarterly,* 29 (1965), pp. 326–337.

FENTON, JOHN H., *Politics in the Border States.* New Orleans: 1957.

GALLI, GIORGIO, and ALFONSON PRANDI, *Patterns of Political Participation in Italy.* New Haven, Conn.: 1970. Covers 1946–1963.

GILPATRICK, THOMAS V., "Price Support Policy and the Midwest Farm Vote," *Midwest Journal of Political Science,* 3 (1959), pp. 319–335.

GLANTZ, OSCAR, "Protestant and Catholic Voting in a Metropolitan Area," *Public Opinion Quarterly,* 23 (1960), pp. 73–82.

GORDON, GLEN, and PHILIP COULTER, "The Sociological Bases of Party Competition: The Case of Massachusetts," *Sociological Quarterly,* 10 (1969), pp. 84–105. Uses multiple regression and factor analysis.

GOSNELL, HAROLD F., *Grass Roots Politics.* Washington: 1942. Correlational analysis by counties.

GOSNELL, HAROLD F., *Machine Politics: Chicago Model.* Chicago: 1937, 1968 ed.). Methodologically innovative (factor analysis).

HACKNEY, SHELDON, *Populism to Progressivism in Alabama.* Princeton, N.J.: 1969.

HENDERSON, HERBERT J., "Political Factions in the Continental Congress: 1774–1783," Unpublished Ph.D. dissertation, Columbia University, 1962.

HOLMES, JACK E., *Politics in New Mexico.* Albuquerque, N.M.: 1967. Uses election returns and roll calls.

HOFFERBERT, R. I., "Classification of American State Party Systems," *Journal of Politics,* 26 (1964), pp. 550–567. Reviews and criticizes past approaches to indexes of inter-party competition.

HOLT, MICHAEL, *Forging a Majority*: Pittsburgh, 1848–1860. New Haven: 1969.

HUDSON, MICHAEL C., "The Electoral Process and Political Development in Lebanon," *Middle East Journal,* 20 (1966), pp. 173–186.

JANOWITZ, MORRIS, and D. R. SEGAL, "Social Cleavage and Party Affiliation: Germany, Great Britain, and the U.S." *American Journal of Sociology,* 72 (1967), pp. 601–618. AID analysis of survey data.

JENNINGS, M. KENT, and HARMON ZEIGLER, "Class, Party, and Race in Four Types of Elections," *Journal of Politics,* 28 (1966), pp. 391–407.

KELLEY, STANLEY, et al., "Registration and Voting: Putting First Things First," *American Political Science Review,* 61 (1967), pp. 359–379.

KESSEL, JOHN, *The Goldwater Coalition.* Indianapolis: 1968.

KESSELMAN, MARK, "French Local Politics: A Statistical Examination of Grass Roots Consensus," *American Political Science Review,* 60 (1966), pp. 963–973.

KEY, O., JR., *American State Politics: An Introduction.* New York: 1956.

———, *Public Opinion and American Democracy.* New York: 1961.

———, *The Responsible Electorate: Rationality in Presidential Voting, 1936–1960.* Cambridge: 1966. Uses poll data to show that majority of voters who switched parties between elections had strong policy reasons for so doing.

————, *Southern Politics in State and Nation.* New York: 1949. Classic.

————, "A Theory of Critical Elections," *Journal of Politics,* 17 (1955), pp. 3–18. Classic.

KINNEAR, MICHAEL, *The British Voter: An Atlas and Survey Since 1885.* Ithaca, N.Y.: 1968.

KLEPPNER, PAUL J., "Lincoln and the Immigrant Vote: A Case of Religious Polarization," *Mid-America,* 48 (1966), pp. 176–195.

KLEPPNER, PAUL, *The Cross of Culture.* New York: 1970. Late nineteenth-century Midwestern politics.

LANCELOT, ALAIN, *L'Abstentionisme electoral en France.* Paris: 1968. Covers 1876–1965.

LEWIS, PIERCE F., "Impact of Negro Migration on the Electoral Geography of Flint, Michigan, 1932–1962: A Cartographic Analysis," *Annals of the Association of American Geographers,* 55 (1965), pp. 1–25.

LIPSET, SEYMOUR and STEIN ROKKAN, eds., *Party Systems and Voter Alignments.* New York: 1967.

MACRAE, DUNCAN, and J. A. MELDRUM, "Critical Elections in Illinois: 1888–1958," *American Political Science Review,* 54 (1960), pp. 669–683. Unsatisfactory attempt to use factor analysis of county returns.

NEPRASH, JERRY, *The Brookhart Campaigns in Iowa 1920–26.* New York: 1932. Correlational analysis; last product of Giddings tradition of electoral geography at Columbia University.

NORTON, JAMES A., "Referenda Voting in a Metropolitan Area," *Western Political Quarterly,* 16 (1963), pp. 195–212.

PETRAS, JAMES F., and MAURICE ZEITLIN, "Miners and Agrarian Radicalism," *American Sociology Review,* 32 (1967), pp. 578–586.

POOL, ITHIEL DE SOLA, ROBERT P. ABELSON, and SAMUEL POPKIN, *Candidates, Issues, and Strategies: A Computer Simulation of the 1960 Presidential Election.* Cambridge: 1964.

POMPER, GERALD, *Elections in America.* New York: 1969. Uses voting statistics, content analysis.

RICE, STUART A., *Farmers and Workers in American Politics.* New York: 1924.

ROBERTS, MICHAEL C., and K. W. RUMAGE, "Spatial Variations in Urban Left-Wing Voting in England and Wales in 1951," *Annals of the Association of American Geographers,* 55 (1965), pp. 161–178.

ROGIN, MICHAEL, *The Intellectuals and McCarthy: The Radical Specter.* Cambridge, Mass.: 1967. Uses scatter diagrams, correlations, and factor analysis, but relies most heavily upon the ecological fallacy.

ROGIN, MICHAEL, and JOHN SHOVER, *Political Change in California*, Westport, Conn.: 1970. Covers 1890–1966.

ROKKAN, STEIN, et al., *Citizens, Elections, Parties*. New York: 1970. Comparative.

ROWNEY, DON KARL, and JAMES Q. GRAHAM, eds., *Quantitative History*. 1969. Part 6.

SANDERSON, G. N., "Swing of the Pendulum in British General Elections, 1832–1966," *Political Studies*, 14 (1966), pp. 349–360.

SEGAL, DAVID R., "Classes, Strata and Parties in West Germany and the U.S.," *Comparative Studies of Sociology and History*, 10 (1967), pp. 66–84. A.I.D. analysis of survey data.

SELLERS, C., "The Equilibrium Cycle in Two Party Politics," *Public Opinion Quarterly*, 29 (1965), pp. 16–38.

STAMMER, OTTO, ed., *Party Systems, Party Organizations and the Politics of the New Masses*. Berlin: 1968.

STOKES, DONALD E., "A Variance Components Model of Political Effects," *Mathematical Applications in Political Science*, vol. 1. Dallas: 1965.

TINGSTEN, HERBERT L., *Political Behavior*. London: 1937.

TURNER, FREDERICK JACKSON, *The Significance of Sections in American History*. New York: 1932. Especially chapters 6 and 11.

———, *The United States: 1830–1850*. New York: 1935.

WEIBULL, JORGEN, "The Wisconsin Progressives, 1900–1914," *Mid-America*, 47 (1965), pp. 190–221.

WYMAN, ROGER E., "Wisconsin Ethnic Groups and the Election of 1890," *Wisconsin Magazine of History*, 51 (1968), pp. 269–293.

4. Legislatures and Roll Calls

ALEXANDER, THOMAS B., *Sec tional Stress and Party Strength: A Study of Roll-Call Voting Patterns in the United States House of Representatives, 1836–1860*. Nashville: 1967. Classic.

ALKER, HAYWARD R., and BRUCE M. RUSSETT, *World Politics in the General Assembly*. New Haven, Conn.: 1965. Factor analysis of roll call votes.

ALLEN, HOWARD W., "Republican Reformers and Foreign Policy, 1913–1917," *Mid-America*, 44 (1962), pp. 222–229.

ANDERSON, LEE F., "Individuality in Voting in Congress: A Research Note," *Midwest Journal of Political Science*, 8 (1964), pp. 425–429.

ANDERSON, LEE F., MEREDITH W. WATTS, JR., and ALLEN R. WILCOX, *Legislative Roll-Call Analysis*. Evanston, Ill.: 1966.

ANDRAIN, CHARLES F., "A Scale Analysis of Senators' Attitudes Toward Civil Rights," *Western Political Quarterly*, 17 (1964), pp. 488–503.

AYDELOTTE, WILLIAM O., "Voting Patterns in the British House of Commons in the 1840s," *Comparative Studies in Society and History*, 5 (1963), pp. 134–163. Collective biography.

BARBER, JAMES, *The Lawmakers*. New Haven, Conn.: 1965. Typology of legislative behavior in Connecticut.

BELKNAP, GEORGE, "A Method for Analyzing Legislative Behavior," *Midwest Journal of Political Science*, 2 (1958), pp. 377–402. Scaling.

BOGUE, ALLAN G., "Bloc and Party in the U.S. Senate, 1861–1863," *Civil War History*, 13 (1967), pp. 221–241. Guttman Scales.

BRUNTON, DOUGLAS, and DONALD H. PENNINGTON, *Members of the Long Parliament*. London: 1954.

CHERRYHOLMES, CLEO and MICHAEL SHAPIRO, *Representatives and Roll Calls: A Computer Simulation of Voting in the Eighty-Eighth Congress*. Indianapolis: 1969.

CLAUSEN, AAGE R., "Measurement Indentity in the Longitudinal Analysis of Legislative Voting," *American Political Science Review*, 61 (1967), pp. 1020–1035. On linking scales for different years.

CLUBB, JEROME M. and HOWARD W. ALLEN, "Party Loyalty in the Progressive Years: The Senate, 1909–1915," *Journal of Politics*, 29 (1967), pp. 567–584.

CNUDDE, C. F., and D. J. MCCRONE, "Linkage between Constituency Attitudes and Congressional Voting Behavior: A Causal Model," *American Political Science Review*, 60 (1966), pp. 66–72.

CRANE, WILDER, "A Caveat on Roll-Call Studies of Party Voting," *Midwest Journal of Political Science*, 4 (1960), pp. 237–249.

FLENN, THOROS, A., and HAROLD L. WALMAN, "Constituency and Roll Call Voting: The Case of Southern Democratic Congressmen," *Midwest Journal of Politics*, 10 (1966), pp. 192–199.

GRAY, CHARLES H., "Scale Analysis of the Voting Records of Senators Kennedy, Johnson, and Goldwater, 1957–1960," *American Political Science Review*, 59 (1965), pp. 615–621. Suggestive.

GROENNINGS, S., W. E. KELLEY, and M. LEISERSON, eds., *The Study of Coalition Behavior*. New York: 1969. Applications of game theory to legislative history.

GROSSMAN, JOEL, and JOSEPH TANENHAUS, eds., *Frontiers of Judicial Research*. New York: 1969.

GRUMM, JOHN G., "A Factor Analysis of Legislative Behavior," *Midwest Journal of Political Science*, 7 (1963), pp. 336–357.

GRUMM, JOHN G., "Systematic Analysis of Blocs in the Study of Legislative Behavior," *Western Political Quarterly*, 18 (1965), pp. 350–362. Cluster analysis.

JEWELL, MALCOLM, and SAMUEL PATTERSON, *The Legislative Process in the United States*. New York: 1966. Fair textbook.

JONES, CHARLES, and RANDALL RIPLEY, *Role of Political Parties in Congress: A Bibliography and Research Guide*. Tucson, Ariz.: 1966.

KOFMEHL, KENNETH, "The Institutionalization of a Voting Bloc," *Western Political Quarterly*, 17 (1964), pp. 256–272.

LIBBY, ORIN G., *Geographical Distribution of the Vote on the Constitution*. Madison, Wisc.: 1894.

———, "Political Factions in Washington's Administration," *Quarterly Journal* (University of North Dakota), 3 (1913), pp. 291–318.

———, "A Plea for the Study of Votes in Congress," American Historical Association *Annual Report for 1896* (Washington: 1897), pp. 321–334.

LINDEN, GLENN, " 'Radicals' and Economic Policies: The House of Representatives, 1861–1873," *Civil War History*, 13 (1967), pp. 51–65.

LINDEN, G. M., " 'Radicals' and Economic Policies: the Senate, 1861–1873," *Journal of Southern History*, 32 (1966), pp. 189–199.

LOWELL, A. LAWRENCE, "The Influence of Party Upon Legislation in England and America," American Historical Association *Annual Report for 1901* (Washington: 1902), pp. 321–542.

MACRAE, DUNCAN, *Parliaments, Parties, and Society in France, 1946–1958*. New York: 1967.

———, "Cluster Analysis of Congressional Votes with the BC TRY System," *Western Political Quarterly*, 19 (1966), pp. 631–638.

———, *Dimensions of Congressional Voting: A Statistical Study of the House of Representatives in the Eighty-First Congress*, University of California Publications in Sociology and Social Institutions. (Vol. I.: 1958).

———, "IBM 1401 Z-Matrix and Editing Programs for Legislative Votes," *Behavioral Science*, 10 (1965), pp. 324–325.

———, "A Method for Identifying Issues and Factions from Legislative Votes," *American Political Science Review*, 59 (1965), pp. 909–926. Basic guide to MacRae's shortcut approach to Guttman scaling. Very important.

———, *Issues and Parties in Legislative Voting*. New York: 1970. Advanced text.

———, and H. D. PRICE, "Scale Positions and 'Power' in the Senate," *Behavioral Science*, 4 (1959), pp. 212–218. A reply to Robert Dahl, "The Concept of Power," (1957), reprinted in Polsby-Dentler-Smith, *Politics and Social Life*. Boston: 1963. Uses early form of success scores.

MAXWELL, GERALD, "Party, Region and the Dimensions of Conflict in the House of Representatives, 1949–1954," *American Political Science Review*, 61 (1967), pp. 380–399.

MAYHEW, DAVID, *Party Loyalty Among Congressmen.* Cambridge: 1966.

MELLER, NORMAN, "Legislative Behavior Revisited: A Review of Five Year's of Publications," *Western Political Quarterly*, 18 (1965), pp. 776–793.

MOORE, JOHN ROBERT, "Conservative Coalition in the U.S. Senate, 1942–1945," *Journal of Southern History*, 33 (1967), pp. 368–376.

MUNGER, F., and J. BLACKHURST, "Factionalism in the National Conventions 1940–1964: An Analysis of Ideological Consistency in State Delegation Voting," *Journal of Politics*, 27 (1965), pp. 375–394.

PRICE, H. DOUGLAS, "Are Southern Democrats Different? An Application of Scale Analysis to Senate Voting Patterns," in Nelson W. Polsby, Robert A. Dentler, and Paul A. Smith, eds., *Politics and Social Life: An Introduction to Political Behavior*. Boston: 1963, pp. 740–756. The remainder of this big anthology is also valuable, especially the essays by Dahl.

RIESELBACH, LEROY, *The Roots of Isolationism: Congressional Voting and Presidential Leadership in Foreign Policy*. Indianapolis: 1966. Roll calls from 76th, 80th, 83rd, and 85th Congresses.

ROACH, HANNAH, "Sectionalism in Congress 1870–1890," *American Political Science Review*, 19 (1925), pp. 500–526.

ROWNEY, DON KARL, and JAMES A GRAHAM, eds., *Quantitative History*. Homewood, Ill.: 1969. Part 2.

SCHUBERT, GLENDON, *Constitutional Politics: The Political Behavior of Supreme Court Justices and the Constitutional Policies that They Make*. New York: 1960.

SCHUBERT, G., "Jackson's Judicial Philosophy: An Exploration in Value Analysis," *American Political Science Review*, 59 (1965), pp. 940–963. Advanced statistical analysis of one judge's votes.

SCHUBERT, GLENDON, ed., *Judicial Behavior*. New York: 1965. Very important.

SCHUBERT, GLENDON, *The Judicial Mind: The Attitudes and Ideologies of Supreme Court Justices 1946–1963*. Evanston, Ill.: 1965. Uses factor analysis and scaling.

SCHUBERT, GLENDON A., *Quantitative Analysis of Judicial Behavior*. Glencoe, Ill.: 1959. Uses cluster bloc, scalograms and game theory.

SHANNON, W. WAYNE, *Party, Constituency, and Congressional Voting*. Baton Rouge, La.: 1969. U.S. House of Representatives, 1959–1961.

SHOVER, JOHN L., "Populism in the Nineteen-Thirties: The Battle for the AAA," *Agricultural History*, 39 (1965), pp. 17–24.

SILBEY, JOEL, *The Shrine of Party.* Pittsburgh: 1967.

TRUMAN, DAVID B., *The Congressional Party: A Case Study.* New York: 1959.

TURNER, JULIUS, *Party and Constituency: Pressures on Congress.* Baltimore: 1970. Covers 1921–1967; illuminating.

ULMER, S. SIDNEY, "Sub-group Formation in the Constitutional Convention," *Midwest Journal of Political Science*, 10 (1966), pp. 288–303.

ULMER, S. S., "Toward a Theory of Sub-Group Formation in the United States Supreme Court," *Journal of Politics*, 27 (1965), pp. 133–152. Uses new clustering techniques.

WAHLKE, JOHN C., and HEINZ EULAU, eds. *Legislative Behavior: A Reader in Theory and Research.* Glencoe, Ill.: 1959. Important.

WAHLKE, JOHN C., HEINZ EULAU, WILLIAM BUCHANAN, and LEROY FERGUSON. *The Legislative System.* New York: 1962.

WOLFF, GERALD, "The Slavocracy and the Homestead Problem of 1854," *Agricultural History*, 40 (1966), pp. 101–111.

WOLFINGER, RAYMOND E., and JOAN HEIFETZ, "Safe Seats, Seniority, and Power in Congress," *American Political Science Review*, 59 (1965), pp. 337–349.

YOUNG, J., *The Washington Community: 1800–1828.* New York: 1966. Imaginative use of residential patterns and roll calls.

5. Collective Biography

ALLEN, HOWARD W., and JEROME M. CLUBB, "Party Loyalty in the Progressive Years: The Senate, 1909–1915," *Journal of Politics*, 29 (1967), pp. 567–584.

AYDELOTTE, WILLIAM O., "Country Gentlemen and the Repeal of the Corn Laws," *English Historical Review*, 82 (1967), pp. 47–60. Collective biography; important.

BERINGER, RICHARD E., "Profile of the Members of the Confederate Congress," *Journal of Southern History*, 33 (1967), pp. 518–541.

BRADLEY, DONALD S., and M. M. ZALD, "From Commercial Elite to Political Administrator: The Recruitment of the Mayors of Chicago," *American Journal of Sociology*, 71 (1965), pp. 153–167.

BRAIBANTI, RALPH, *Asian Bureaucratic Systems Emergent from the British Imperial Tradition.* Durham, N.C.: 1966. Collective biography 1600–1960.

CHANG, CHUNG-LI, *The Chinese Gentry: Studies on Their Role in Nineteenth-Century Chinese Society.* Seattle: 1955.

CLARKE, EDWIN, *American Men of Letters: Their Nature and Nurture.* New York: 1916. Pioneering statistical study.

DOLLARD, JOHN, *Criteria for the Life History*. Gloucester, Mass.: 1949.

DONALD, DAVID, *The Politics of Reconstruction, 1863–1867*. Baton Rouge, La.: 1965.

EDINGER, L. J., "Political Science and Political Biography: Reflections on the Study of Leadership," *Journal of Politics*, 26 (1964), pp. 423–439; pp. 648–675.

EDINGER, LEWIS J., and D. D. SEARING, "Social Background in Elite Analysis: A Methodological Inquiry," *American Political Science Review*, 61 (1967), pp. 428–445.

FRYKENBERG, ROBERT, "Elite Groups in a South Indian District: 1788–1858," *Journal of Asian Studies*, 24 (1965), pp. 261–281.

GATELL, FRANK OTTO, "Money and Party in Jacksonian America: A Quantitative Look at New York City's Men of Quality," *Political Science Quarterly*, 82 (1967), pp. 235–252.

GEHLEN, MICHAEL P., and MICHAEL MCBRIDE, "The Soviet Central Committee: An Elite Analysis," *American Political Science Review*, 62 (1968), pp. 1232–1241.

GRUNER, ERICH, "Freiheit and Bindung in den Bundesratswahlen," in *Annuaire Suisse de Science Politique* (Lausanne: 1967). Thorough, quantitative study of recruitment to the Swiss Federal Council, 1848–1966.

JENSEN, RICHARD, "Quantitative Collective Biography," in Robert Swierenga, ed., *Quantification in American History*. New York: 1970, pp. 389–405. New computerized methods.

KEELER, MARY F., *The Long Parliament, 1640–41: A Biographical Study of its Members*. Philadelphia: 1954.

KNIGHT, MAXWELL E., *The German Executive, 1890–1933*. Stanford: 1955. Cabinet members.

KUBOTA, AKIVA, *Higher Civil Servants in Post-War Japan*. Princeton, N.J.: 1969.

LASSWELL, HAROLD D., DANIEL LERNER, and C. EASTON ROTHWELL, *The Comparative Study of Elites*. Stanford, Calif.: 1952.

LLOYD, P. C., ed., *The New Elites of Tropical Africa*. Oxford: 1966.

LODGE, MILTON, "Soviet Elite Participatory Attitudes in the Post-Stalin Period," *American Political Science Review*, 62 (1968), pp. 827–839. Content analysis.

MAIN, JACKSON TURNER, *The Upper House in Revolutionary America 1763–1788*. Madison: 1967.

MATTHEWS, DONALD R., *U.S. Senators and their World*. Chapel Hill, N.C.: 1960. Important.

MILLER, WILLIAM, ed., *Men in Business: Essays in the History of Entrepreneurship*. Cambridge, Mass.: 1952.

NEALE, JOHN E., *The Elizabethan House of Commons.* London: 1949.

⸻, *Elizabeth I and Her Parliaments.* London: 1953.

PRITCHETT, C. HERMAN, *The Roosevelt Court: A Study in Judicial Politics, 1937–1947.* New York: 1948.

RABB, THEODORE K., *Empire and Enterprise: Investment in English Overseas Enterprise 1575–1630.* Cambridge: 1968. Computerized.

RIGBY, T. H., *Communist Party Membership in the U.S.S.R., 1917–1967.* Princeton, N.J.: 1968.

ROWNEY, DON KARL and JAMES Q. GRAHAM, *Quantitative History* Homewood, Ill.: 1969. Part 2.

RUSTOW, D. A., "Study of Elites: Who's Who, When, and How," *World Politics,* 18 (1966), pp. 690–717. Review essay.

SCHLESINGER, JOSEPH A., *Ambition and Politics: Political Careers in the United States.* Chicago: 1967.

SCHMIDT, HANNELORE, "Die Deutsche Exekutive, 1949–1960," *Archives Européennes de Sociologie,* 4 (1965), pp. 166–176.

SINGER, M. R., *The Emerging Elite: A Study of Political Leadership in Ceylon.* Cambridge: 1964.

STAUFFER, ROBERT B., "Philippine Legislators and Their Changing Universe," *Journal of Politics,* 28 (1966), pp. 556–597.

STONE, LAWRENCE, *The Crisis of the Aristocracy, 1558–1641.* Oxford: 1965.

WOOSTER, RALPH, *The People in Power.* Knoxville, Tenn.: 1969. Covers Southern Officeholders in 1850 and 1860.

ZAPF, WOLFGANG, *Wandlungen der deutschen elite: ein Zirkulations modell deutschen Führungsgruppen, 1919–1961.* Munich: 1965. With useful bibliography.

ZEMSKY, ROBERT, "Power, Influence, and Status: Leadership Patterns in the Massachusetts Assembly, 1740–1755," *William and Mary Quarterly,* 26 (1969), pp. 502–520.

C. SOCIAL AND ECONOMIC

1. Economic History

ADELMAN, IRMA, "Long Cycles—Fact or Artifact?", *American Economic Review,* 55 (1965), pp. 444–463.

ADELMAN, IRMA and CYNTHIA TAFT MORRIS, *Society, Politics and Economic Development.* Baltimore: 1967. Extensive use of factor analysis.

BAILYN, BERNARD, and LOTTE BAILYN, *Massachusetts Shipping: A Statistical Study.* Cambridge: 1959.

BJORK, G., "The Weaning of the American Economy: Independence, Market Changes, and Economic Development," *Journal of Economic History,* 24 (1964), pp. 541–560. On 1780s.

BLYN, GEORGE, *Agricultural Trends in India, 1891–1947*. Philadelphia: 1966.

BRAUDEL, FERNAND, *La Méditerranée et le Monde Méditerranéen a l'Epoque de Philippe II*, 2 vols. Paris; English translation: London: 1949, 1966. Imaginative use of quantitative data in cultural study.

CAMERON, RONDO, *Banking in the Early Stages of Industrialization*. New York: 1967. For England, Scotland, France, Belgium, Germany, Russia, and Japan.

CIPOLLA, CARLO M., *Money, Prices, and Civilization in the Mediterranean World, Fifth to Seventeenth Century*. Princeton, N.J.: 1956.

CLARK, COLIN, *The Conditions of Economic Progress*, 3d rev. ed. New York: 1957. Useful data.

COCHRAN, THOMAS C., "Did the Civil War Retard Industrialization?", *Mississippi Valley Historical Review*, 48 (1961), pp. 197–210.

COHEN, JON S., "Financing Industrialization in Italy, 1894–1914: The Partial Transformation of a Late-Comer," *Journal of Economic History*, 27 (1967), pp. 363–382.

Conference on Research in Income and Wealth, *Output, Employment, and Productivity in the United States After 1800*. New York: 1966. Important NBER collection of studies.

CONRAD, ALFRED H., and JOHN R. MEYER, *The Economics of Slavery and Other Essays in Econometric History*. Chicago: 1964.

DAVID, PAUL A., "Growth of Real Product in the U.S. before 1840: New evidence, Controlled Conjectures," *Journal of Economic History*, 27 (1967), pp. 151–197.

DAVIES, GEORGE, "Social Aspects of the Business Cycle," *Quarterly Journal* (U. of North Dakota), 12 (1922), pp. 105–121.

DAVIS, LANCE E., and JOHN LEGLER, "Government in the American Economy, 1815–1902: A Quantitative Study," (with reply by D. McDougall), *Journal of Economic History*, 26 (1966), pp. 514–555. Multiple regression analysis including time as variable.

DEANE, PHYLLIS, and W. A. COLE, *British Economic Growth: 1688–1959*. Cambridge: 1962, Sophisticated.

EASTERLIN, RICHARD A., "Economic-Demographic Interactions and Long Swings in Economic Growth," *American Economic Review*, 56 (1966), pp. 1063–1064.

FABRICANT, SOLOMON, *The Trend of Government Activity in the United States since 1900*. New York: 1952.

FISHLOW, ALBERT, *American Railroads and the Transformation of the Antebellum Economy*. Cambridge: 1965.

FOGEL, ROBERT W., "New Economic History: Its Findings and Methods," *Economic History Review*, 19 (1966), pp. 642–663.

FOGEL, ROBERT, *Railroads and American Economic Growth*. Baltimore: 1964. Classic about canals that never existed.

FRIEDMAN, MILTON, and ANNA JACOBSON SCHWARTZ. *A Monetary History of the United States*, 1867–1970. Princeton, N.J.: 1963. Classic in American economic history.

GENTIL DA SILVA, JOSÉ, *En Espagne, Developpement Economique, Subsistance, Declin*. Paris: 1965. Spanish economic and demographic history; quantitative.

GREENHUT, MELVIN, and W. TATE WHITMAN, eds., *Essays in Southern Economic Development*. Chapel Hill, N.C.: 1964.

HAZLEWOOD, ARTHUR, *The Economics of Development*. London: 1964. Annotated bibliography of books and articles published 1958–1962. Very useful.

HENDERSON, JAMES, and ANNE KRUEGER, *National Growth and Economic Change in the Upper Midwest*. Minneapolis: 1965.

HOOLEY, R. W., "The Measurement of Capital Formation in Underdeveloped Countries," *Review of Economics and Statistics*, 49 (1967), pp. 199–208. Points to a large systematic error over time.

HOSELITZ, B. F., "On Historical Comparisons in the Study of Economic Growth," in National Bureau of Economic Research, *The Comparative Study of Economic Growth and Structure*. New York: 1959, pp. 145–161.

HUNT, EDWARD H., "New Economic History: Professor Fogel's Study of American Railways," *History*, 53 (1968), pp. 3–18. With reply by G. R. Hawke.

International Association for Research in Income and Wealth, *Bibliography on Income and Wealth*, 8 vols. Cambridge, England: 1952. A continuing series.

KEYFITZ, N., "Age Distribution as a Challenge to Development," *American Journal of Sociology*, 70 (1965), pp. 659–668. Reply with rejoinder by van de Walle, E., 71 (1966), pp. 549–557.

KUZNETS, S., *Modern Economic Growth: Rate, Structure, and Spread*. New Haven: 1966. Major work.

KUZNETS, SIMON, *National Income and Its Composition*, 1919–1938, 2 vols. New York: 1941.

LABROUSSE, CAMILLE E., *La crise de l'économie francaise à la fin de l'ancien régime et au début de le Révolution*. Paris: 1944.

LANE, FREDERIC C., and JELLE C. RIEMERSMA, eds., *Enterprise and Secular Change*. Homewood, Ill.: 1953.

LEONTIEF, W. W., "The Structure of the U.S. Economy," *Scientific American*, 212 (April, 1965), pp. 25–35.

LORENZ, CHARLOTTE, *Forschungslehre der Sozialstatistik*, 3 vols. Berlin: 1951–1963.

MCCLELLAND, E. M., "Railroads, American Growth, and the New Economic History: A Critique," *Journal of Economic History,* 28 (1967), pp. 102–123.

MADDISON, ANGUS, *Economic Growth in the West: Comparative Experience in Europe and North America.* New York: 1964. Valuable for data.

MARCZEWSKI, J., ed., *Histoire quatitative de l'économie francaise,* v. 3 of F. Perroux, ed., *Cahiers de l'Institut de Science Economique Appliquée.* Paris: 1963. Marczewski is the leading French cleometricien.

MITCHELL, WESLEY C., *Business Cycles, the Problem and its Setting.* New York: 1927.

MOORSTEEN, RICHARD and RAYMOND POWELL, *The Soviet Capital Stock, 1928–1962.* Homewood, Ill.: 1966.

MORRIS, M. DAVID, and BURTON STEIN, "The Economic History of India: A Bibliographic Essay," *Journal of Economic History,* 21 (1961), pp. 179–207. See also Morris, "Towards a Reinterpretation of 19th Century Indian Economic History," 23 (1963), pp. 608–618.

MUKHERJEE, M., "A Preliminary Study of Growth of National Income in India, 1857–1957," in Int. Assn. for Research in Income and Wealth, *Asian Studies in Income and Wealth.* Bombay: 1965, pp. 71–103.

MURPHY, GEORGE, "The 'New' History," *Explorations in Entrepreneurial History,* 2 (1965), pp. 132–146. Critical.

NEISSER, HANS, and FRANCO MODIGLIANI, *National Incomes and International Trade: A Quantitative Analysis.* Urbana, Ill.: 1953.

NORTH, DOUGLASS, *The Economic Growth of the United States: 1790–1860.* Englewood Cliffs, N.J.: 1961.

NORTH, DOUGLASS, *Growth and Welfare in the American Past.* Englewood Cliffs, N.J.: 1966.

PATEL, S. J., "The Economic Distance between Nations: Its Origin, Measurement and Outlook," *Economic Journal,* 74 (1964), pp. 119–131.

PERLOFF, HARVEY, et al., *Regions, Resources and Economic Growth.* Baltimore: 1960. Valuable for American economic regionalism since 1870.

Purdue Faculty Papers in Economic History, Homewood, Ill.: 1967. Important.

ROSTOW, W. W., ed., *The Economics of Take-off into Sustained Growth.* London: 1963. Important symposium. Good review and interpretation of growth data for developed countries; destroyed the Rostow thesis.

ROWNEY, DON KARL, and JAMES Q. GRAHAM, eds., *Quantitative History.* Homewood, Ill.: 1969. Part 5.

SANCHEZ RAMOS, FRANCISCO, *La Económica Sidúrgica Española.* Madrid: 1945. Economic history twentieth century Spain; important statistical series.

SCHUMPETER, JOSEPH A., *Business Cycles,* 2 vols. New York: 1939.

SCHWEITZER, ARTHUR, "Economic Systems and Economic History," *Journal of Economic History,* 25 (1965), pp. 660–679.

Studies in Income and Wealth. Washington, 1947. A continuing series of volumes, most of them conference reports, on U.S. National Income. Basic to economic historians.

SVENNILSON, INGVAR, *Growth and Stagnation in the European Economy.* Geneva: 1954.

TEMIN, PETER, "In Pursuit of the Exact," *Times Literary Supplement* (1966), pp. 652–653. Interesting.

TEMIN, PETER, "Labor Scarcity and the Problem of American Industrial Efficiency in the 1850s," *Journal of Economic History,* 26 (1966), pp. 277–298. Reply by Ian Drummond, 27 (1967), pp. 383–390.

THOMAS, ROBERT P., "Quantitative Approach to the Study of the Effects of British Imperial Policy upon Colonial Welfare; Some Preliminary Findings," *Journal of Economic History,* 25 (1965), pp. 615–638.

TOSTLEBE, ALVIN S., *Capital in Agriculture: Its Formation and Financing Since 1870.* New York: 1957.

TUNZELMANN, G. N. VON, "The New Economic History: An Econometric Appraisal," *Explorations in Entrepreneurial History,* 5 (1968), pp. 175–200.

VARTANIAN, PERSHING, "The Cochran Thesis: A Critique in Statistical Analysis," *Journal of American History,* 51 (1964), pp. 77–89.

VICENS VIVES, JAIME, and J. NADAL, *Manual de Historia Económica de España.* Barcelona: 1959. Important bibliographic guide.

VITORINO DE MAGALHAIS GODINHO, *Prix et Monnaies au Portugal, 1750–1850.* Paris: 1955. Important source on this particular topic; in series Monnaie-Prix-Conjuncture, ed. Lucien Febvre et al.

WISH, JOHN R., *Economic Development in Latin America: An Annotated Bibliography.* New York: 1965. Broad in scope, but less useful than Hazlewood.

WOLMAN, LEO, *Ebb and Flow in Trade Unionism.* New York: 1936.

2. Education and Human Capital

ABEL, J. F., and N. J. BOND, *Illiteracy in the Several Countries of the World.* Washington: 1929.

ANDERSON, C. A., and M. J. BOWMAN, eds., *Education and Economic Development.* Chicago: 1965.

BECKER, GARY, *Human Capital.* Princeton, N.J.: 1964. Basic.

BECKER, GARY, *Economics of Discrimination*. Chicago: 1957. Important.

BLAUG, MARK, *Economics of Education: A Selected Annotated Bibliography*. New York and Oxford: 1966.

CAMPBELL, ROBERT, and BARRY N. SIEGEL, "The Demand for Higher Education in the United States, 1919–1964," *American Economic Review*, 57 (1967), pp. 482–494. Econometric.

COLBERG, M. R., *Human Capital in Southern Development, 1939–1963*. Chapel Hill, N.C.: 1965.

DENISON, EDWARD, *Why Growth Rates Differ*. Washington: 1967. Major comparative analysis of post-World War II economic growth in Europe and U.S., with stress on human capital.

FOLGER, JOHN K., and CHARLES NAM, *Education of the American Population*. Washington: 1967. Important census monograph.

JANSEN, MARIUS B., and LAWRENCE STONE, "Education and Modernization in Japan and England," *Comparative Studies in Society and History*, 9 (1967), pp. 208–232.

KATZ, ELIHU, et al., "Traditions of Research on the Diffusion of Innovations," *American Sociological Review*, 28 (1963), pp. 237–252.

LEBERGOTT, STANLEY, *Manpower in Economic Growth: The American Record Since 1800*. New York: 1964.

MORSE, D., and A. W. WARNER, eds., *Technological Innovation and Society*. New York: 1966.

PELZ, D. C., and F. M. ANDREWS, *Scientists in Organizations: Productive Climates for Research and Development*. New York: 1966. Uses multivariate statistical methods.

STONE, L., "The Educational Revolution in England, 1560–1640," *Past and Present*, 37 (1964), pp. 41–80.

———, "Literacy and Education in England 1640–1900," *Past and Present*, 42 (1969), pp. 69–139.

3. Demography and Population

ADELMAN, I., and C. T. MORRIS, "A Quantitative Study of Social and Political Determinants of Fertility," *Economic Developments and Cultural Change*, 14 (1966), pp. 129–157. Factor analysis of demographic indices for 55 poor countries.

BELOCH, KARL J., *Bevölkerungsgeschichte Italiens*, Gaetano de Sanctis, ed., 2 vols. Berlin and Leipzig: 1937–1939.

BELOCH, KARL J., *Die Bevölkerung der griechisch-römischen Welt*. Leipzig: 1886.

BESHERS, JAMES M., *Population Processes in Social Systems*. New York: 1967. Uses newer quantitative methods; historical time series.

BOGUE, DONALD J., *The Population of the United States*. Glencoe, Ill.: 1959.

BOGUE, DONALD, and CALVIN BEALE, *Economic Areas of the United States*. New York: 1961. Important statistical compilation; divides each state into "natural" economic regions and describes each.

BOGUE, DONALD J., *Principles of Demography*. New York: 1969. The standard text; contains much valuable data and superb bibliographies. To be supplemented by Donald Bogue and Evelyn Kitagawa, *A Manual of Demographic Research Techniques* (in press), with a thorough discussion of numerical techniques.

CARR-SAUNDERS, A. M., *World Population: Past Growth & Present Trends*. London: 1936. Basic.

CLARKE, JOHN, *Population Geography*. New York: 1965. Good brief introduction to field.

CURTIN, PHILIP D., *The Atlantic Slave Trade: A Census*. Madison, Wisc.: 1969.

DAVIS, KINGSLEY, *The Population of India and Pakistan*. Princeton, N.J.: 1951.

DRAKE, MICHAEL, *Population and Society in Norway, 1735–1865*. Cambridge, England: 1969.

ELDRIDGE, HOPE T., *The Materials of Demography; a Selected and Annotated Bibliography*. New York: 1959. Important guide.

FRUMKIN, GREGORY, *Population Changes in Europe Since 1939: A Study of Population Changes in Europe During and Since World War II as shown by the Balance Sheets of Twenty-four European Countries*. New York: 1951.

GLASS, DAVID V., *Population Policies & Movements in Europe*, 2d ed., Oxford: 1940, 1967. Useful.

GLASS, D. V., and D. E. C. EVERSLEY, eds., *Population in History*. Chicago: 1965. Important anthology of larger articles on Europe and U.S. in seventeenth, eighteenth, and nineteenth centuries.

GREVEN, PHILIP J., *Four Generations: Population, Land, and Family in Colonial Andover, Massachusetts*. Ithaca: 1970.

HABAKKUK, H. J., and M. M. POSTAN, eds., *Cambridge Economic History of Europe*. Cambridge: 1966. Volume 6 contains several important summary articles and detailed guides to census reports. The other volumes are also valuable.

HASKETT, RICHARD C., "An Introductory Bibliography for the History of American Immigration, 1607–1955"; Lavell, Carr' B., and Wilson E. Schmidt, "An Annotated Bibliography on the Demographic, Economic and Sociological Aspects of Immigration," in George Washington University; *A Report on World Population Migrations as Related to the United States of America*. Washington: 1956.

HATT, PAUL, and ALBERT REISS, eds., *Cities and Society.* Glencoe, Ill.: 1957. Important reader on urban sociology.

HAUSER, P. M., and O. D. DUNCAN, eds., *The Study of Population: An Inventory and Appraisal.* Chicago: 1959.

HENRY, LOUIS, *Manuel de démographie historique.* Geneva, Paris: 1967. For French techniques and studying family structures.

HIGGS, ROBERT, and H. LOUIS STETTLER, "Colonial New England Demography: A Sampling Approach," *William and Mary Quarterly,* 27 (1970), pp. 282–294.

HOLLINGSWORTH, T. H., *Historical Demography.* London: 1969. Review of research; full bibliography.

HUTCHINSON, EDWARD P., *Immigrants and Their Children: 1850–1950.* New York: 1956. A census monograph.

International Encyclopedia of the Social Sciences. New York: 1968. Articles on Population; Migration; Demography; Cities; Census.

KELLEY, ALLEN, "International Migration and Economic Growth in Australia: 1865–1965," *Journal of Economic History,* 25 (1965), pp. 333–354.

KUCZYNSKI, ROBERT R., *Demographic Survey of the British Colonial Empire,* 3 vols. London: 1948–1953.

KUZNETS, SIMON, *Modern Economic Growth.* New Haven: 1966. Relates nineteenth and twentieth century demographic factors to economic growth. Important for theory.

——— and ERNEST RUBIN, *Immigration and the Foreign Born.* New York: 1954. Notes about 40 percent of the immigrants to the U.S., 1890–1910, returned to their homeland.

——— and DOROTHY THOMAS, eds., *Population Redistribution and Economic Growth,* 3 vols. Philadelphia: 1957–1964. Major study of U.S., 1870–1950, with many valuable new series of migration, wealth and income by states.

LEE, EVERETT and ANNE, "Internal Migration Statistics for the United States," *Journal of American Statistical Association,* 55 (1960), pp. 664–697. On state-of-birth data in federal census, and migration reports 1940.

LEGEARD, C., *Guide de recherches documentaires en démographie* Paris: 1966. The basic guide to the field, to primary sources, bibliography and techniques.

LEVASSEUR, EMILE, *La population française: histoire de la population avant 1789 et démographie de la France comparée à celle des autres nations au XIX siècle, précédée d'une introduction sur la statistique,* 3 vols. Paris: 1889–1892. Classic.

MOLS, ROGER, *Introduction à la démographie historique des villes d'Europe, du XIVe au XVIIIe siècle,* 2 vols. Louvain: 1954–1956.

MONKHOUSE, F. J., and H. R. WILKINSON, *Maps and Diagrams.* New York: 1963. Important chapter on mapping population distributions.

NADEL, JORGE, *La Población Española*. Barcelona: 1966.

ODUM, HOWARD W., and HARRY E. MOORE, *American Regionalism*. New York: 1938. Multidisciplinary.

RAZZELL, P. E., "Population Change in 18th-Century England: A Reinterpretation," *Economic Historical Review*, 18 (1965), pp. 312–332.

REINHARD, MARCEL, "Demography, the Economy, and the French Revolution," in Evelyn Acomb, ed., *French Society and Culture since the Old Regime*. New York: 1966, pp. 20–42. See also his "Autour de la Révolution française," in *Annales*, 21 (1966), pp. 201–226.

REINHARD, MARCEL, and ANDRÉ ARMENGAUD, *Histoire Générale de la Population Mondiale*. Paris: 1961.

REVELLE, ROGER, ed., "Historical Population Studies," *Daedalus*, 97 (1968), pp. 353–635. Excellent guide to current state of the field.

ROGERS, ANDREI, and ROBERT MILLER, "Estimating a Matrix Population Growth Operator from Distributional Time Series," *Annals of the Association of American Geographers*, 57 (1967), pp. 751–756.

ROGERS, ANDREI, *Matrix Analysis of Interregional Population Growth and Distribution*. Berkeley: 1968.

ROSENBLAT, ANGEL, *La población indígena y el mestizaje en América*, 2 vols. Buenos Aires: 1954.

ROWNEY, DON KARL, and JAMES Q. GRAHAM, *Quantitative History*. Homewood, Ill.: 1969. Part 4.

RUSSELL, JOSIAH C., *Late Ancient and Medieval Population*. Philadelphia: 1958.

SIEGEL, JACOB, and MELVIN ZELNIK, "An Evaluation of Coverage in the 1960 Census Population by Techniques of Demographic Analysis and by Composite Methods," American Statistical Association, *Proceedings of the Social Statistics Section*, 1966, (Washington: 1966), pp. 71–85.

SMITH, C. T., *An Historical Geography of Western Europe Before 1800*. New York: 1967. Concerns population distribution and movements, and economic patterns.

SMITH, T. L., "Redistribution of the Negro Population of the U.S. 1910–1960." *Journal of Negro History*, 51 (1966), pp. 155–173.

SMITH, T. LYNN, *Fundamentals of Population Study*. Philadelphia: 1967. Good undergrad textbook.

SPENGLER, JOSEPH J., and OTIS D. DUNCAN, eds., *Demographic Analysis*. Glencoe, Ill.: 1956. Very useful anthology for research techniques, population theory and policy, and selected readings.

SUTHERLAND, STELLA, *Population Distribution in Colonial America*. New York: 1936.

TAEUBER, CONRAD and IRENE, *The Changing Population of the U.S.* New York: 1958. Census study.

THEODORSON, G. A., ed., *Studies in Human Ecology.* Evanston, Ill.: 1961. Important.

THOMAS, BRINLEY, *Migration and Economic Growth: A Study of Great Britain and the Atlantic Economy.* Cambridge, England: 1954.

———, ed., *Economics of International Migration.* New York: 1958.

THOMAS, DOROTHY S., *Social and Economic Aspects of Swedish Population Movements, 1750–1933.* New York: 1941.

THOMPSON, WARREN, and P. K. WHELPTON, *Population Trends in the U.S.* New York: 1933. Old but very valuable for historical trends.

United Nations, Department of Social Affairs, Population Division, *The Determinants and Consequences of Population Trends: A Summary of the Findings of Studies on the Relationships between Population Changes and Economic and Social Conditions.* Population Studies #17. New York: 1953. Old, but still useful.

———, Methods of Estimating Basic Demographic Measures from Incomplete Data. Population Studies #42. New York: 1967.

United Nations, Statistical Office. *Demographic Yearbook,* Annual, (1949). Basic compilation of current data; continues Statistical Yearbook of the League of Nations. Geneva: 1927–1944.

WHITNEY, HERBERT A., "Estimating Precensus Populations: A Method Suggested and Applied to the Towns of Rhode Island and Plymouth Colonies in 1689," *Annals of the Association of American Geographers,* 55 (1965), pp. 179–189.

WILLCOX, WALTER, ed., *International Migrations,* 2 vols. New York: 1929, 1931. Basic source of data and interpretation.

WOYTINSKI, W., *Die Welt in Zahlen,* 7 vols. Berlin: 1928. Vol. 1 has excellent summary of population distributions in the early 1920s.

WOYTINSKI, WLADIMIR and EMMA, *World Population and Production.* New York: 1953. Good for data c. 1950.

WRIGLEY, E. A., ed., *An Introduction to English Historical Demography.* London: 1966. Disappointingly narrow. Long bibliography for England.

WRIGLEY, E. A., *Population and History.* New York: 1969. A good introduction to historical demography.

ZELINSKY, WILBUR, *A Bibliographic Guide to Population Geography.* Chicago: 1962. 2588 items.

4. Urbanization and Community

BEDARIDA, FRANCOIS, "Londres au milieu de XIXᵉ siècle: une analyse de structure sociale," *Annales*, 23 (1968), pp. 268–295.

CHECKLAND, S. G., *The Rise of Industrial Society in England, 1815–1885.* New York: 1964.

CONVERSE, PHILIP E., "The Nature of Belief Systems in Mass Publics," in David E. Apter, ed., *Ideology and Discontent.* New York: 1964, pp. 206–261.

CURTI, MERLE, *The Making of An American Community.* Stanford: 1959. Reconstruction of mid-nineteenth century social, political, and economic structures of a rural Wisconsin county.

DEUTSCH, KARL, W., *Nationalism and Social Communication*, rev. ed. Cambridge: 1965.

GIBBS, J. B., "Measures of Urbanization," *Social Forces*, 45 (1966), pp. 170–177.

HAUSER, P. M., and L. F. SCHNORE, eds., *The Study of Urbanization.* New York: 1965. Important, interdisciplinary, good on history.

HAWLEY, A. H., *Human Ecology.* New York: 1950.

HOFFERBERT, RICHARD, "Socioeconomic Dimensions of the American States: 1890–1960," *Midwest Journal of Political Science*, 12 (1968), pp. 401–418.

JANSON, CARL-GUNNAR, "The Spatial Structure of Newark, New Jersey: Part 1, The Central City," *Acta Sociologica*, 11 (1968), pp. 144–169. Imaginative use of factor analysis.

JOHNSTON, NORMAN J., "The Caste and Class of the Urban Form of Historic Philadelphia," *Journal of the American Institute of Planners* (1966), pp. 334–349. Distribution of church membership before Civil War.

JONES, E., "The Distribution and Segregation of Roman Catholics in Belfast," *Sociological Review*, 4 (1956), pp. 167–189.

MOORE, W., *The Impact of Industry.* Englewood Cliffs, N.J.: 1965. Historical sociology.

MORRILL, RICHARD, "The Development of Spatial Distributions of Towns in Sweden: A Historical-Predictive Approach," *Annals of the Association of American Geographers*, 53 (1963), pp. 1–14.

MORRILL, RICHARD, "The Negro Ghetto: Problems and Alternatives," *Geography Review*, 55 (1965), pp. 339–361. Computer simulation of historic process, Seattle, 1940–1960.

MOSER, CLAUS, *British Towns: A Statistical Study of their Social and Economic Difference.* Edinburgh: 1961.

ORTIZ, DOMINGUEZ, *La Sociedad Española en el Siglo, XVIII.* Madrid: 1955. *Lo Sociedad Española en el Siglo XVII.* Madrid: 1963. Basic studies of Spanish social history, with guide to statistical sources.

OSTROWSKI, KRZYSTOF, and ADAM PRZEWORSKI, "Preliminary Inquiry into the Nature of Social Change: The Case of the Polish Countryside," *International Journal of Comparative Sociology,* 8 (1967), pp. 26–43.

SCHMID, CALVIN, *Social Trends in Seattle.* Seattle: 1944.

SOROKIN, PITIRIM A., et al., eds., *A Systematic Source Book in Rural Sociology,* 3 vols. Minneapolis: 1930–1932.

THERNSTOM, STEPHAN, and RICHARD SENNETT, eds., *Nineteenth Century Cities.* New Haven: 1969. Important.

TILLY, CHARLES, "State of Urbanization: Review Article," *Comparative Studies in Society and History,* 10 (1967), pp. 100–113.

WARNER, SAM BASS, "If All the World Were Philadelphia," *American Historical Review,* 74 (1968), pp. 26–43.

WILLIAMSON, JEFFREY G., "Antebellum Urbanization in the American Northeast," *Journal of Economic History,* 25 (1965), pp. 592–608.

YOUNG, PAULINE, *Scientific Social Surveys and Research.* New York: 1956. Guide to community studies.

5. Mobility, Conflict, Stratification, and Organizations

AL-SAMARRIE, AHMAD, and H. P. MILLER, "State Differentials in Income Concentration," *American Economic Review,* 57 (1967), pp. 59–72. Multiple regressions to explain Gini indices.

ANDIC, S., and A. T. PEACOCK, "The International Distribution of Income, 1949 and 1957," *Journals of the Royal Statistical Society,* 124 (1961), pp. 206–218.

BAER, W., *Regional Inequality and Economic Growth in Brazil,* Economic Development and Cultural Change, 12 (1964), pp. 268–285.

BENDIX, R., and S. LIPSET, eds., *Class, Status and Power: Social Stratification in Comparative Perspective.* New York: 1966.

BLALOCK, HUBERT M., *Toward a Theory of Minority-Group Relations.* New York: 1967. Model building on ecological data.

BLAU, PETER, and OTIS DUDLEY DUNCAN, *American Occupational Structure.* New York: 1967. The methodological classic of the 1960s.

BLOOMBAUM, MILTON, "The Conditions Underlying Race Riots as Portrayed by Multidimensional Scalogram Analysis," *American Sociological Review,* 33 (1968), pp. 76–89.

BLUMIN, STUART, "The Historical Study of Vertical Mobility," *Historical Methods Newsletter,* 1 (1968), pp. 1–13.

BWY, DOUGLAS, "Dimensions of Social Conflict in Latin America," *American Behavioral Scientist,* 11 (1968), pp. 39–50.

CAPECCHI, V., "Problèms méthodologiques dans la mesure de la mobilité social," *Archives européennes de sociologie,* 8 (1967), pp. 285–318.

CHAMBERLAYNE, J., "From Sect to Church in British Methodism," *British Journal of Sociology*, 15 (1964), pp. 139–149.

COOPER, J. P., "Social Distribution of Land and Men in England, 1436–1700," *Economic History Review*, 20 (1967), pp. 419–440.

CUTRIGHT, PHILLIPS, "Inequality: A Cross-National Analysis," *American Sociology Review*, 32 (1967), pp. 562–578.

DAVISSON, WILLIAM, "Essex County Wealth Trends: Wealth and Economic Growth in Seventeenth Century Massachusetts," *Essex Institute Historical Collections*, 103 (1967), pp. 317–325.

DYE, THOMAS R., "Urban School Segregation: A Comparative Analysis," *Urban Affairs Quarterly*, 4 (1969), pp. 141–166.

DUOCASTELLA, ROGELIO, "Géographie de la Practique Religieuse en Espagne," *Social Compass (International Review of Socio-Religious Studies)*, 12 (1965), pp. 253–302. For historical sociology and data series on Spanish Church.

GLENN, NORVAL D., and J. L. SIMMONS, "Are Regional Cultural Differences Diminishing?", *Public Opinion Quarterly*, 31 (1967), pp. 176–193.

GOODMAN, LEO A., "On the Statistical Analysis of Mobility Tables," *American Journal of Sociology*, 70 (1965), pp. 564–585.

GRAHAM, HUGH DAVIS, and TED GURR, eds., *Violence in America: Historical and Comparative Perspectives*. Washington: 1969.

GURR, TED R., "Psychological Factors in Civil Violence," *World Politics* 20 (1968), pp. 245–278. "Urban Disorder: Perspectives from the Comparative Study of Civil Strife," *American Behavioral Scientist*, 11 (1968), pp. 50–55. "A Causal Model of Civil Strife," *American Political Science Review*, 62 (1968), pp. 1104–1124.

———, *Why Men Rebel*. Princeton: 1970.

HACKNEY, SHELDON, "Southern Violence," *American Historical Review*, 74 (1969), pp. 906–925.

HENRETTA, JAMES A., "Economic Development and Social Structure in Colonial Boston," *William and Mary Quarterly*, 22 (1965), pp. 75–92.

HO, PING-TI, *The Ladder of Success in Imperial China: Aspects of Social Mobility, 1368–1911*. New York: 1962. Collective biography.

HSU, C., *Ancient China in Transition*. Stanford, Calif.: 1965. Social mobility 722–222 B. C.

JACKSON, J. A., ed., *Social Stratification*. Cambridge: 1968. Comparative.

KAPLOW, JEFFRY, *Elbeuf during the Revolutionary Period: History and Social Structure*. Baltimore: 1964.

KATZ, D., and R. KAHN, *The Social Psychology of Organizations*. New York: 1966. Important synthesis.

KOCH, DONALD WARNER, "Income Distribution and Political Structure in Seventeenth Century Salem, Massachusetts," *Essex Institute Historical Collections*, 105 (1969), pp. 49–71.

KOVAL'CHENKO, I. D., "Analysis of Historical Statistics by Mathematical Means," *Soviet Studies in History*, 3 (1964), pp. 13–20.

LADURIE, EMANUAL LE ROY, *Les paysans de Languedoc*. Paris: 1966.

LASUEN, J. R., "Regional Income Inequalities and the Problems of Growth in Spain," *Papers of the Regional Science Association, European Congress*, 8 (1962), pp. 169–191.

LEON, PIERRE, ed., *Structures économiques et problèmes sociaux du monde rural dans la France du Sud-Est: fin du XVIIe siècle–1835*. Paris: 1966.

LIPSET, SEYMOUR MARTIN and REINHARD BENDIX, *Social Mobility in Industrial Society*. Berkeley, Calif.: 1959.

LUPSHA, PETER A., "On Theories of Urban Violence," *Urban Affairs Quarterly*, 4 (1969), pp. 273–296.

MAIN, JACKSON TURNER, *The Social Structure of Revolutionary America*. Princeton, N.J.: 1965.

MARCH, JAMES G., ed., *Handbook of Organizations*. Chicago: 1965. Important, comprehensive.

MILLER, HERMAN P., *Income Distribution in the United States*. Washington: 1966. Major collection of data.

OWSLEY, FRANK, *Plain Folk of the Old South*. Baton Rouge, La.: 1949. Summarizes census studies by Owsley and his students.

SCHELLING, T. C., *Models of Segregation*. Santa Monica, Calif.: 1969. RAND study RM-6014-RC.

SCHOLFIELD, R. S., "Geographical Distribution of Wealth in England, 1334–1649," *Economic History Review*, 18 (1965), pp. 483–510.

SMELSER, NEIL, and S. M. LIPSET, eds., *Social Structure and Mobility in Economic Development*. Chicago: 1966. Important, especially the essay on methodology by O. D. Duncan.

SOROKIN, PITIRIM A., *Social Mobility*. New York: 1927.

STONE, LAWRENCE, "Social Mobility in England, 1500–1700," *Past and Present*, 33 (1967), pp. 16–55.

TAEUBER, KARL and ALMA, *Negroes in Cities*. Chicago: 1965. Standard source on measurement of segregation in U.S., with historical data and important methodological chapter.

THERNSTROM, STEPHEN, *Poverty and Progress: Social Mobility in a Nineteenth Century City*. Cambridge: 1964.

THERNSTROM, STEPHEN, "Notes on the Historical Study of Social Mobility," *Comparative Studies in Society and History*, 10 (1968).

THOMPSON, I. A. A., "A Map of Crime in Sixteenth-Century Spain," *Economic History Review*, 21 (1968), pp. 244–267. Relates distribution of galley oarsmen to social and economic conditions.

THOMPSON, F. J. L., "Social Distribution of Landed Property in England Since the Sixteenth Century," *Economic History Review*, 19 (1966), pp. 505–517.

TILLY, CHARLES, *The Vendée*. Cambridge: 1964.

TUMIN, MELVIN, *Social Stratification*. Englewood Cliffs, N.J.: 1967. Useful introduction.

Appendix A: Functions of Percentages

P	$1-P$	$P(1-P)$	$\sqrt{P(1-P)}$	$1-P^2$	$1-(1-P)^2$	$\sqrt{1-p^2}$	$\sqrt{1-(1-p)^2}$
.00	1.00	.0000	.00000	1.0000	.0000	1.0000	.0000
.01	.99	.0099	.09950	.9999	.0199	.9999	.1411
.02	.98	.0196	.14000	.9996	.0396	.9998	.1990
.03	.97	.0291	.17059	.9991	.0591	.9995	.2431
.04	.96	.0384	.19596	.9984	.0784	.9992	.2800
.05	.95	.0475	.21794	.9975	.0975	.9987	.3122
.06	.94	.0564	.23749	.9964	.1164	.9982	.3412
.07	.93	.0651	.25515	.9951	.1351	.9975	.3676
.08	.92	.0736	.27129	.9936	.1536	.9968	.3919
.09	.91	.0819	.28618	.9919	·.1719	.9959	.4146
.10	.90	.0900	.30000	.9900	.1900	.9950	.4359
.11	.89	.0979	.31289	.9879	.2079	.9939	.4560
.12	.88	.1056	.32496	.9856	.2256	.9928	.4750
.13	.87	.1131	.33630	.9831	.2431	.9915	.4931
.14	.86	.1204	.34699	.9804	.2604	.9902	.5103
.15	.85	.1275	.35707	.9775	.2775	.9887	.5268
.16	.84	.1344	.36661	.9744	.2944	.9871	.5426
.17	.83	.1411	.37563	.9711	.3111	.9854	.5578
.18	.82	.1476	.38419	.9676	.3276	.9837	.5724
.19	.81	.1539	.39230	.9639	.3439	.9818	.5864
.20	.80	.1600	.40000	.9600	.3600	.9798	.6000
.21	.79	.1659	.40731	.9559	.3759	.9777	.6131
.22	.78	.1716	.41425	.9516	.3916	.9755	.6258
.23	.77	.1771	.42083	.9471	.4071	.9732	.6380
.24	.76	.1824	.42708	.9424	.4224	.9708	.6499
.25	.75	.1875	.43301	.9375	.4375	.9682	.6614
.26	.74	.1924	.43863	.9324	.4524	.9656	.6726
.27	.73	.1971	.44396	.9271	.4671	.9629	.6834
.28	.72	.2016	.44900	.9216	.4816	.9600	.6940
.29	.71	.2059	.45376	.9159	.4959	.9570	.7042
.30	.70	.2100	.45826	.9100	.5100	.9539	.7141
.31	.69	.2139	.46249	.9039	.5239	.9507	.7238
.32	.68	.2176	.46648	.8976	.5376	.9474	.7332
.33	.67	.2211	.47021	.8911	.5511	.9440	.7424
.34	.66	.2244	.47371	.8844	.5644	.9404	.7513
.35	.65	.2275	.47697	.8775	.5775	.9367	.7599
.36	.64	.2304	.48000	.8704	.5904	.9330	.7684
.37	.63	.2331	.48280	.8631	.6031	.9290	.7766
.38	.62	.2356	.48539	.8556	.6156	.9250	.7846
.39	.61	.2379	.48775	.8479	.6279	.9208	.7924
.40	.60	.2400	.48990	.8400	.6400	.9165	.8000
.41	.59	.2419	.49183	.8319	.6519	.9121	.8074
.42	.58	.2436	.49356	.8236	.6636	.9075	.8146
.43	.57	.2451	.49508	.8151	.6751	.9028	.8216
.44	.56	.2464	.49639	.8064	.6864	.8980	.8285
.45	.55	.2475	.49749	.7975	.6975	.8930	.8352
.46	.54	.2484	.49840	.7884	.7084	.8879	.8417
.47	.53	.2491	.49910	.7791	.7191	.8827	.8480
.48	.52	.2496	.49960	.7696	.7296	.8773	.8542
.49	.51	.2499	.49990	.7599	.7399	.8717	.8602
.50	.50	.2500	.50000	.7500	.7500	.8660	.8660

Appendix B: Squares and Values for the Computation of Spearman Rho, r_s

D	D^2	D	D^2	N	$\dfrac{N(N^2-1)}{6}$
½	¼	20½	420¼		
1	1	21	441	3	4
1½	2¼	21½	462¼	4	10
2	4	22	484	5	20
2½	6¼	22½	506¼	6	35
3	9	23	529	7	56
3½	12¼	23½	552¼	8	84
4	16	24	576	9	120
4½	20¼	24½	600¼	10	165
5	25	25	625	11	220
5½	30¼	25½	650¼	12	286
6	36	26	676	13	364
6½	42¼	26½	702¼	14	455
7	49	27	729	15	560
7½	56¼	27½	756¼	16	680
8	64	28	784	17	816
8½	72¼	28½	812¼	18	969
9	81	29	841	19	1140
9½	90¼	29½	870¼	20	1330
10	100	30	900	21	1540
10½	110¼	30½	930¼	22	1771
11	121	31	961	23	2024
11½	132¼	31½	992¼	24	2300
12	144	32	1024	25	2600
12½	156¼	32½	1056¼	26	2925
13	169	33	1089	27	3276
13½	182¼	33½	1122¼	28	3654
14	196	34	1156	29	4060
14½	210¼	34½	1190¼	30	4495
15	225	35	1225	31	4960
15½	240¼	35½	1260¼	32	5456
16	256	36	1296	33	5984
16½	272¼	36½	1332¼	34	6545
17	289	37	1369	35	7140
17½	306¼	37½	1406¼	36	7770
18	324	38	1444	37	8436
18½	342¼	38½	1482¼	38	9139
19	361	39	1521	39	9880
19½	380¼	39½	1560¼	40	10660
20	400	40	1600		

Appendix C: Nomograph for Spearman rho, r_s

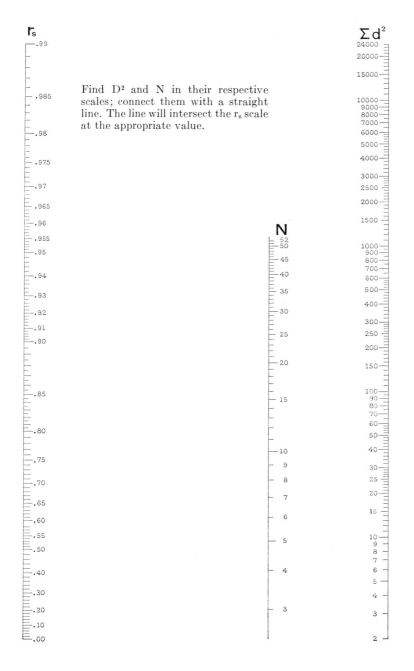

r_s

- .99
- .985
- .98
- .975
- .97
- .965
- .96
- .955
- .95
- .94
- .93
- .92
- .91
- .90
- .85
- .80
- .75
- .70
- .65
- .60
- .55
- .50
- .40
- .30
- .20
- .10
- .00

Find D^2 and N in their respective
scales; connect them with a straight
line. The line will intersect the r_s scale
at the appropriate value.

N

- 52
- 50
- 45
- 40
- 35
- 30
- 25
- 20
- 15
- 10
- 9
- 8
- 7
- 6
- 5
- 4
- 3

Σd^2

- 24000
- 20000
- 15000
- 10000
- 9000
- 8000
- 7000
- 6000
- 5000
- 4000
- 3000
- 2500
- 2000
- 1500
- 1000
- 900
- 800
- 700
- 600
- 500
- 400
- 300
- 250
- 200
- 150
- 100
- 90
- 80
- 70
- 60
- 50
- 40
- 30
- 25
- 20
- 15
- 10
- 9
- 8
- 7
- 6
- 5
- 4
- 3
- 2

Source: Jack Dunlap and Albert Kurtz, *Handbook of Statistical Nomographs, Tables, and
Formulas* (Yonkers-on-Hudson, 1932) p. 37.

INDEX

THE INDEX BELOW WAS REPRODUCED DIRECTLY FROM THE PRINTOUT OUT OF AN IBM 360 SERIES MODEL 65 COMPUTER AT OKLAHOMA STATE UNIVERSITY. KEYWORD INDEXING TECHNIQUES SIMILAR TO THOSE DESCRIBED IN CHAPTER 6 WERE EMPLOYED. THE INDEXING PROGRAM WAS DEVELOPED BY MR. WILLIAM V. ACCOLA III AND MR. DAVID PICKFRING OF THE OKLAHOMA STATE UNIVERSITY COMPUTER CENTER STAFF. TO USE THE INDEX, SEARCH THE CENTER COLUMN OF ALPHABETIZED WORDS FOR THE DESIRED ENTRY. PAGE NUMBERS ARE GIVEN AT THE END OF EACH LINE.